1986

ELECTRONIC POST-PRODUCTION

THE FILM-TO-VIDEO GUIDE

Gary H. Anderson

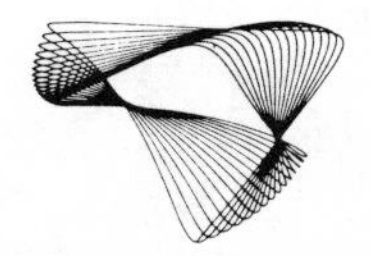

Knowledge Industry Publications, Inc.
White Plains, NY and London

Video Bookshelf

Electronic Post-Production: The Film-to-Video Guide

Library of Congress Cataloging-in-Publication Data
Anderson, Gary H.
Electronic post-production: the film-to-video guide.

Bibliography: p.
Includes index.
1. Moving-pictures—Editing. 2. Video tapes—Editing. I. Title.
TR899.A55 1986 778.5′235 86-10255
ISBN 0-86729-155-9

Printed in the United States of America

10 9 8 7 6 5 4 3 2 1

Contents

List of Tables and Figures

Acknowledgments

I am deeply indebted to Philip Miller for another excellent effort in transforming my manuscript into this book and to the following companies for photos and data: Eastman Kodak Co.; Grass Valley Group, Inc.; Lucasfilm Ltd.; Marconi Broadcast Division, The Marconi Co. Ltd.; Montage Computer Corp. and Rank Cintel Ltd.

I also want to express my sincerest thanks to Mr. Roy Brubaker for his invaluable advice and expertise, as well as Steve Buchsbaum, David Cipes, June Galas, Beth Hagen, Pam Malouf, Mike Matheson, Frank Merwald, Ed Rowan, Oley Sassone, Simon Straker and Andy Zall.

Finally I want to express my deepest gratitude to my wife, Diane, and her invaluable word-processing expertise, and to my sons, Darrell and Craig, for their patience and encouragement throughout the preparation of this manuscript.

Introduction

If a brochure from the Eastman Kodak Co. is to be believed, in the motion picture industry of the 1990s, "the only place you will find scissors and splicers is in the studio museum."[1]

Kodak's claim may overstate the situation somewhat; it does, however, highlight an important industry trend: The electronic post-production of film material has been increasing rapidly, and it will continue to increase in the coming years.

Consider these revealing facts. According to figures published by the Society of Motion Picture Technicians and Engineers (SMPTE), the use of film as a production medium was up 17% in the first half of 1984 compared to 1983. However, whereas negative film sales were also up during that period, 16mm print sales actually decreased. This is due in considerable part to the increased use of electronic post-production techniques, resulting in less need for film workprints and final release prints.[2]

DEVELOPMENT OF ELECTRONIC POST-PRODUCTION

For years, video producers and filmmakers have experimented with editing feature films using video techniques. Due to the limitations in video technology, however, their early efforts met with results that ranged from mediocre to poor.

In one technique, feature films were shot on video, edited electronically and then transferred to film. Unfortunately, this resulted in a rather low-resolution final product because of the limited capability of the 525 scanline television picture (see Chapter 3). Another technique involved shooting the feature on film, transferring the film material to videotape, editing the videotape electronically and then transferring the edited program back to film—with the same low-resolution results.

For feature film projects, there were also several attempts at creating workprints on videotape, electronically editing the video workprints and then using the edited workprints to generate the cutting list needed to conform (perform the final editing on) the film negative. To aid in this process, computer software programs were developed to convert the videotape time code back to film edge numbers. In addition, film transfer systems were designed to insert edge number information into the time code used in electronic editing. In several of these systems, edge numbers were actually burned into the video workprints, and the editor was required to log them by hand after offline editing. Like other pioneering efforts, these techniques met with widely varying degrees of success.

THE FUTURE OF ELECTRONIC POST-PRODUCTION: FILM VERSUS VIDEO

Until a very high definition television (HDTV) system yielding at least 35mm print quality is developed, or until laser video projection or satellite delivery systems are installed in theaters, feature films will continue to be delivered to theaters as 35mm prints. As a result, the future for the electronic post-production of feature films will continue to rest on the development of a reliable, efficient method for generating a frame-accurate cutting list for editing the film negative, as well as the development of a system for dealing with multitrack sound editing. Fortunately, new technologies such as EditDroid, SoundDroid and Montage Picture Processor (all of which are described in Chapter 5) should help in these areas.

The situation for filmmakers whose project is to be delivered on videotape is far more encouraging. In fact, for several years now, producers of television commercials have routinely generated footage on film first and then transferred the footage to video for final editing (see Chapter 2). Using state-of-the-art video equipment, they can quickly and accurately complete dissolves, picture mortising and the other, more elaborate special effects used so frequently in TV commercials. As I discuss in Chapter 2, the producers of music videos have also been quick to pick up on the advantages of electronic post-production for generating special effects.

When time and budget constraints are a factor, the relatively fast turnaround possible with film-to-video transfer, along with the relatively low cost of offline editing and online conformation, can make electronic post-production an appealing alternative for a wide variety of film projects. In fact, the producers of many network television series that shoot on film are currently taking a close look at the "electronic posting" alternative—particularly with the post-production successes of programs such as "The Paper Chase," "AfterMASH" and "Fame."

DECIDING WHETHER TO "GO ELECTRONIC"

The conditions described above confirm that, as the Eastman Kodak brochure suggests, most post-production of film projects will "go electronic" in the coming years. For filmmakers and television professionals working today, however, many questions remain. Will electronic post-production really save time and money on a given project? What are the benefits? The drawbacks? And what is actually involved in "going electronic?"

In the chapters that follow, I've tried to anticipate and answer the different questions surrounding the electronic post-production of film material. Drawing on my many years of experience as a video editor, I've described the various technical, budgetary and creative considerations involved in determining whether electronic post-production is appropriate for a particular film project. I've also described the different stages of electronic post-production, from film-to-tape transfer to final online conformation. The result is a handbook intended to be both straightforward enough to serve as a primer for newcomers to the field and detailed enough to serve as a reference guide for practicing professionals. In particular, I hope that *Electronic Post-Production: The Film-to-Video Guide* will help film and video professionals answer the one question that many production projects will face in future years: Even though I started my project on film, should I finish it on videotape?

NOTES

1. Eastman Kodak Co., Datakode sales brochure.
2. *Society of Motion Picture Technicians and Engineers Journal* (April 1985): 398.

1 Deciding on Electronic Post-Production

Today, all filmmakers face a fundamental decision as they approach post-production: Should they stay with the traditional "slice and splice" method of film editing, or should they switch to one or more forms of electronic post-production? To make the correct decision, film professionals must consider a number of factors, including the final release format of the production, the scope and complexity of post-production audio requirements and scheduling and budgeting constraints.

In this chapter, I discuss each of these factors, analyzing the different considerations and concerns that producers must take into account as they weigh the various post-production options. After reading the sections that follow, producers should be better prepared to decide for themselves which form of post-production best fits the requirements of their particular projects.

FINAL RELEASE FORMAT

Before deciding whether to go electronic, production professionals should first consider the final release format on which the finished project will be delivered to television stations, movie theaters and other destinations. For productions that will be sent to TV stations, the final release format is almost always videotape—even when the project was shot and edited on film. In contrast, feature films that are shown in movie theaters are currently delivered as 35mm film prints. As discussed in the Introduction, this will continue to be the case until projection television technology reaches, or at least approximates, 35mm quality.

Other projects, including educational and industrial productions, can be delivered on either film or videotape. In recent years, however, the trend has been to videotape distribution for these nonbroadcast projects, particularly as more schools and corporate playback sites have become equipped with ½-inch Beta and VHS videotape decks.

Switching to electronic post-production makes the most sense when the final delivery format is videotape. Making the switch allows you to take advantage of the many benefits that electronic editing offers, and you end up with a videotape master that is ready for duplication and distribution.

Of course, even when the final delivery is film, electronic post-production can still be the best solution for many projects. This is particularly true when the production faces time and budgetary constraints—conditions discussed later in this chapter.

POST-PRODUCTION SOUND REQUIREMENTS

The core of film sound is the ability to overlap dialog tracks, to edit multiple sync and effects tracks at frame and subframe intervals, and to control each track separately. Historically, most electronic post-production projects didn't need this level of complexity nor was early state-of-the-art electronics capable of providing it. Consequently, electronic post-production lacked many of the technical abilities film producers required.

In electronic post-production, editors currently choose from one of several methods to prepare and synchronize (sync) the sound track. At present, the most common method is to transfer the ¼-inch nagra sound track recorded during shooting to 35mm magnetic film, which is then synced (synchronized) with the dailies for double (video and sound) system transfer to videotape.

Another method involves transferring the ¼-inch nagra track directly to videotape and then syncing the film negative with the sound track during the film-to-tape transfer stage. In addition, music video projects often premix the final sound for production playback, while commercial productions will often slice, splice and mix the magnetic film in the conventional way and then sync the final sound track with the finished film workprint during film-to-tape transfer.

Although each of these methods has proven effective on certain types of projects, film producers continue to be concerned about the problem of overlapping sound tracks and "slipping" individual tracks during electronic post-production.

Overlapping Sound Tracks

The problem of overlapping sound tracks during electronic post-production is not difficult to solve—as long as you are using a time code–based video editing system that includes player/recorders capable of two-track sound editing. With this type of system, you can "split track" and extend each sound edit simply by telling the editing computer which track to use for recording.

You can also solve the multiple track and overlapping problems by connecting a 16- or 24-track audio recorder to the computerized editing system during the online conformation (final electronic editing) stage. However, even with state-of-the-art video editing systems, this would require using several workprint copies during offline editing, each containing two sound tracks. You would then end up with separate edit decision lists for each pair of sound tracks.

Slipping Sound Tracks

As mentioned above, the ability to "slip" sound tracks is another major concern for filmmakers who have opted for electronic post-production. Because all of the dialog tracks are usually recorded on one master multitrack audiotape, slipping one track would typically require making a copy of the track, "sliding" it and recording back onto the master multitrack.

One alternative to this technique is to use an additional multitrack audiotape recorder (ATR) that contains an identical recording of the dialog sound tracks, in conjunction with a time code–based editing controller.

Another, more sophisticated method employed at Glen Glenn Sound involves synchronizing the multitrack audio recorder with sets of 35mm dummies—machines used for playing back 35mm magnetic sound tracks. This process is described in more detail in Chapter 6.

As digital sound technology matures, still more options are becoming available. For example, it is now possible to edit dialog or sound effects tracks digitally, recalling sound segments with instant random access, extending or shortening individual bits of sound information (including individual musical notes) and slipping sound tracks with an accuracy of better than one-tenth frame—all with no generational loss of sound quality. In the near future, sound editors and mixers will be able to use computerized post-production systems to retrieve, manipulate, record and edit individual bits of sound information or entire split-track dialog sequences.

Influencing the Decision to "Go Electronic"

Even with all of these present and pending technological alternatives, electronic "posting" remains a more viable alternative on projects with relatively moderate post-production sound requirements. However, technical advances such as the post audio processing system described in Chapter 6 promise to extend the range and sophistication of sound mixing options available during electronic post-production. As a result, in the near future, even projects with extremely complex sound requirements will become legitimate candidates for electronic posting. For more information, see Chapter 6.

SCHEDULING RESTRICTIONS

Remote shoots that are fouled up, network air dates that are suddenly moved up, last-minute script changes and other unanticipated events can raise havoc with even the most carefully planned post-production schedules. In many cases, electronic post-production can help buy some extra time, particularly when the project will ultimately end up on videotape. For this reason, projects with tight delivery schedules should consider electronic post-production, particularly during the opticals and duplication for final delivery stages.

In 1982, Universal Studios and Consolidated Film Industries (CFI) cooperated on a comparison test to determine how much time electronic post-production can save over conventional film editing. For the test, Universal supplied CFI with copies of all the dailies (the film from each day of shooting) for one 45.5 minute episode of "Magnum P.I.," a prime-time television series. CFI was charged with editing the episode on video, while Universal edited it employing normal film techniques.

Using the Rank Cintel telecine machine, CFI first transferred 28,000 feet of dailies onto 28 reels of ¾-inch videocassettes and 1-inch type C videotape. The first cut edit on a computer editing system required approximately 650 edits and took about 69 hours. Ultimately, the final post-production analysis revealed that Universal required a total of 262 hours to edit the episode on film, compared with the 105 hours that CFI required to finish the production electronically—a time saving of more than 166%.[1]

Of course, this test wasn't completely scientific because there were too many uncontrolled variables—for instance, there was no central person in charge of overseeing both projects. In addition, the producers viewed and gave revisions based solely on the film version (the film version was always intended to be aired). Consequently, the video revisions were also based on these notes.

Still, the "Magnum P.I." experiment indicates that, given the right set of conditions, electronic post-production can help save substantial amounts of time.

BUDGETARY CONSIDERATIONS

Film production budgets have risen dramatically in recent years, primarily due to increases in labor costs, in the price of film stock and in the fees and interest on production financing. Unfortunately, efforts to reduce expenses in these key areas have met with only limited success. As a result, many studios and producers have turned their attention to cutting costs in the area of post-production—an area where they have more alternatives and more budgetary options. In fact, by switching to electronic post-production for the TV series "Fame," MGM claims to have saved 20% to 30% over their previous post-production costs.

Labor Costs

Labor costs account for a substantial portion of any post-production budget. Electronic post-production can help reduce those labor costs, primarily because the switch to electronic posting can help save time. For broadcast TV projects, the most significant cost savings will probably come from reducing the number of editing teams working on different episodes at any one time. Given the time savings that can result from switching to electronic post-production, a TV series that was using three editing teams, for instance, could probably meet production deadlines with just two teams.

I would not recommend making drastic or random staffing cuts, however, unless you are prepared to suffer the same fate faced by one weekly cable television series. Assuming that electronic post-production would work miracles, the series' producers budgeted for only one editing team, no sound editors, and no sound effects personnel. The savings in labor costs were more than offset by the trouble created when the production missed its air dates. As this unfortunate example illustrates, electronic posting has not yet reached, and may never reach, the point where producers can safely eliminate key members of the post-production team.

When considering the labor savings possible through electronic post-production, you should also think about the impact of any existing agreements with the various unions and the Directors' Guild. For example, by Guild agreement, directors must be allotted a certain amount of time in post-production before they are assigned to their next project. If electronic posting reduces the amount of time a director must spend in post-production, is the director free to start on a new project sooner than the Guild agreement specifies?

Union agreements also specify the number of below-the-line crew members required on unionized film projects, as well as which crew members are responsible for which production and post-production functions. When electronic posting blurs the lines between the various functions or creates job overlap, who performs a given task? Squabbles between the various unions and locals are already a common occurrence, especially now that the film locals have finally admitted that their video counterparts are here to stay. The only way to solve the squabbling is to create new union guidelines that take the changes brought by electronic posting into account.

Equipment Costs

Electronic post-production systems cost much more ($100,000 and up) than the movieola systems used in conventional film editing. Will these increased costs be balanced by the overall savings that result from switching to electronic posting? The answer is often "yes," particularly if the production company or studio can amortize the cost of the equipment over a 7-to-10-year period, or perhaps even over a shorter period. For instance, MGM has stated that the savings realized by switching to electronic post-production equipment paid for the new equipment in the first year of operation.

Production companies that decide against laying out the full purchase price of an electronic editing system might want to look into negotiating a package post-production deal with an appropriately equipped video facility. All video facilities compete for business, and they are most likely to give the best price when they are handling all of your post-production needs. A package deal can also make life much easier for producers, since all the post-production is done under one roof. In particular, this tends to cut down on the problems and mishaps that result when materials need to be handled by and transported to different facilities.

A word of warning to companies that do plan to invest in a full electronic post-production system. Because the state of the art in video equipment changes so rapidly, be sure to negotiate some sort of upgrade policy with the manufacturer or distributor. Otherwise, you may find yourself stuck with a high-priced system that quickly becomes outdated, putting you at a disadvantage with your more up-to-date competitors.

Materials and Processing Costs

Switching to electronic post-production can also save money in the areas of materials and film processing. Although you would still have to pay for the film negative stock used in shooting the production and any magnetic film needed for dailies, all other film stock costs would be eliminated. Of course, instead of film stock, you would need to purchase videotape stock for film-to-video transfers, workprint cassettes, edited masters and release copies. Also, in both traditional and electronic post-production, you would still need to add in any stock costs required to edit the sound track. Traditional film stock costs would include any ¼-inch audiotape, 35mm magnetic film stock and multitrack audiotape, whereas electronic post-production stock costs would include only the multitrack audiotape and possibly some ¼-inch stock because the sound and picture are edited onto the same videotape stock. When you add it all up, however, going electronic almost always results in considerable savings in materials costs—especially when you figure in any additional fees for processing the film stock.

Admittedly, not all film processing costs are eliminated in electronic post-production. If the project was shot on film, you still must develop the original footage and pick up any charges for syncing the dailies. Then, you must add in the "electronic lab" fees that you will incur in electronic post-production. Those fees include the cost of (1) transferring the film negative to videotape; (2) making videotape workprints; and (3) duplicating the final release copies.

You must also compare the cost of editing (conforming) the final master program on videotape to the cost of cutting the negative in conventional film editing. Finally, you must consider the cost of color timing, opticals and basic sound editing required in film editing. As Table 1.1 and Table 1.2 show, once you've sorted out the variables, electronic post-production is almost always less expensive than conventional film editing—at least as far as materials and processing costs are concerned.

Electronic Post-Production versus Conventional Film Editing: A Comparison Budget

In late 1984, the producers of a weekly, prime-time television series decided to draft a budget that compared the cost of conventional film editing—the method that they were currently using—to the cost of electronic post-production. Here are some vital production factors for the series:

- The series was shot on film.
- Each episode aired in a one-hour time slot.
- Daily production averaged about 6000 feet of film.
- With a seven-day shooting schedule, weekly production averaged 42,000 feet of film.

The comparison budget prepared for the series is shown in Tables 1.1 and 1.2.

As Table 1.1 and Table 1.2 show, the comparison budget indicated that switching to electronic post-production would result in a saving of about $60,000 per episode—even with the added cost of film-to-tape transfer. The producers also estimated that going electronic would have saved them considerable time in post-production.

Unfortunately, the series was canceled before the producers could make the switch. As a result, it is impossible to determine what the exact time and money savings would have been. However, even if the producers were guilty of underestimating some costs and overestimating the time savings, it is clear that going electronic would have been the right choice—at least for this particular project.

Table 1.1: Per-Episode Comparison Budget: Conventional Film Editing (Actual)

Picture editing and sound crew	$34,000
Post-production audio/syncing with dailies	20,000
Stock shots	5,000
Opticals	4,800
Titles	4,500
Lab processing/release print	12,000
Miscellaneous extras	13,000
Total	$93,300

Table 1.2: Per-Episode Comparison Budget: Electronic Post-Production (Estimated)

Film-to-tape transfer	
Footage transferred: 24,000 feet (from a total of 42,000 feet shot)	
Estimated transfer time: 12 hours	
Costs (including labor):	
Film-to-tape transfer (12 hours at $350 per hour)	$4,200
Dailies (240 minutes at $5.25 per minute)	1,260
1-inch tape stock (4.5 hours at $150 per hour)	675
Dubbing of 1-inch master to ¾-inch cassettes for offline editing (4.5 hours at $103 per hour)	464
Transferring stock shots (1 hour at $350 per hour)	350
Stock shots	
Rental, library costs	$5,000
Offline editing	
¾-inch editing (80 hours at $130 per hour)	$10,400
Online conformation	
Online editing using four machines (12 hours at $550 per hour)	$6,600
Preparing titles using CHYRON (2 hours at $150 per hour)	300
1-inch tape stock for master (60 minutes)	150
Audio post-production	
Pre-lay (8 hours at $135 per hour)	$1,080
Sound effects (8 hours at $250 per hour)	2,000
Layback (1 hour at $230 per hour)	230
Distribution	
Dubbing (two 1-inch copies for airing at $135 each)	$270
1-inch tape stock (2 reels at $150 each)	300
Total	$33,279

BUDGETING FOR ELECTRONIC POST-PRODUCTION

Table 1.1 provides a general overview of the costs involved in electronic post-production. This sort of overview is fine for comparing the costs of electronic posting and conventional film editing. However, once you decide to go electronic, you'll need to prepare a much more detailed budget. The sample budget form in Figure 1.1 is designed to help you do just that.

As you can see, the budget form is divided into eight cost categories: laboratory, stock shots, sound, scoring session, special facilities, opticals and animation, video and miscellaneous. In addition, I've included a sample design for what may be the most important component of any post-production budget: the cover page. I've described each of these categories and components in the sections that follow.

As you read about the different budget items, keep this one overriding consideration in mind: No single project will incur all of the costs I've detailed in the sample budget. For example, if you are editing the master on videotape rather than film, you probably won't need the reversal dupes listed as a budget item under laboratory costs. Determining which items apply to your production is the first step in developing an accurate, detailed budget.

Cover Page

The cover page of your budget should serve as ready reference for those involved with and responsible for the post-production session. First, it should contain the names, addresses and phone numbers of key project personnel, including designated contacts for both the production company and the post-production facility. The cover page should also provide vital statistics about the project and the post-production session, including the title, running time and scheduled editing and delivery dates. Finally, and perhaps most important, the cover page should contain a summary of and subtotals for the different cost categories.

Laboratory Costs

Usually, laboratory fees for electronic post-production include the cost of developing and printing the film negative. However, some producers consider this a production item, since the film must be developed before electronic post-production can begin. The costs in this category include the price of developing the negative and, if necessary, printing a "one light" copy (a copy printed using only one exposure setting).

Coding, another item listed under laboratory costs, is the process of printing a series of numbers along the edge of the film and the magnetic (mag) audio track

Figure 1.1: Post-Production Budget Form

DATE__________	PRODUCTION NUMBER__________
PRODUCTION COMPANY__________	POST-PRODUCTION FACILITY__________
ADDRESS__________	ADDRESS__________
TELEPHONE__________	TELEPHONE__________
CONTACT__________	CONTACT__________
DIRECTOR__________	PRODUCER__________
TELEPHONE__________	TELEPHONE__________
__________	__________
__________	__________

PRODUCTION TITLE	RUNNING TIME	ID#
__________	__________	__________
__________	__________	__________
__________	__________	__________

BEGIN EDITING DATE__________	DELIVER FINAL PRODUCT TO__________
FINISH EDITING DATE__________	ADDRESS TO__________
FINISH SOUND DATE__________	TELEPHONE__________
DELIVERY DATE__________	CONTACT__________

<u>COST SUMMARY TOTAL</u>

A. LABORATORY COSTS	__________
B. STOCK SHOT COSTS	__________
C. SOUND COSTS	__________
D. SCORING SESSION COSTS	__________
E. SPECIAL FACILITIES COSTS	__________
F. OPTICALS AND ANIMATION COSTS	__________
G. VIDEO COSTS	__________
H. MISCELLANEOUS COSTS	__________
SUB-TOTAL	__________
SALES TAX (____%)	__________
TOTAL	__________

Figure 1.1: Post-Production Budget Form (Cont.)

CATEGORY	UNITS	RATE	SUB-TOTAL
A. LABORATORY COSTS			
DEVELOPING AND PRINTING	______	______	______
CODING	______	______	______
REPRINTING (35mm/16mm)	______	______	______
REVERSAL DUPES B/W-COLOR	______	______	______
DUPE CRI (35mm/16mm)	______	______	______
NEGATIVE PREP/CONFORM	______	______	______
ANSWER PRINT	______	______	______
16mm REDUCTION ANSWER PRINT	______	______	______
35mm PROTECTION IP	______	______	______
35mm RELEASE PRINTS	______	______	______
		TOTAL FOR CATEGORY	______
B. STOCK SHOT COSTS			
LIBRARY COSTS	______	______	______
LAB PROCESSING	______	______	______
RENTAL	______	______	______
		TOTAL FOR CATEGORY	______
C. SOUND COSTS			
NARRATION RECORDING	______	______	______
ADR SESSION	______	______	______
SOUND EFFECTS RECORDING	______	______	______
STOCK MUSIC SEARCH/FEE	______	______	______
TEMPORARY MIX	______	______	______
SWEETENING AND MIXING	______	______	______
MAG TRANSFERS	______	______	______
LAUGH MACHINE	______	______	______
OPTICAL SOUND FILM	______	______	______
STOCK COSTS	______	______	______
LABOR	______	______	______
		TOTAL FOR CATEGORY	______

Figure 1.1: Post-Production Budget Form (Cont.)

CATEGORY	UNITS	RATE	SUB-TOTAL
D. SCORING SESSION COSTS			
RIGHTS AND LICENSES			
LYRICIST			
COMPOSER			
ARRANGER			
COPYIST AND PRINTING			
ORCHESTRATOR			
MUSICIANS			
INSTRUMENT RENTAL			
INSTRUMENT HAULING			
VIDEO/FILM PLAYBACK EQUIPMENT			
PROJECTIONIST			
SCORING STAGE			
STAGE CREW			
MUSIC EDITOR			
MIXDOWN			
		TOTAL FOR CATEGORY	
E. SPECIAL FACILITIES COSTS			
SCREENING ROOM			
EQUIPMENT RENTAL			
LABOR			
		TOTAL FOR CATEGORY	
F. OPTICALS AND ANIMATION COSTS			
ARTWORK			
TITLE PHOTOGRAPHY			
MATTE PHOTOGRAPHY			
HiCON DEVELOP AND PRINTING			
OPTICAL EFFECTS			
LAB PROCESSING			
AMORTIZATION OF TITLE WORK			
ANIMATOR AND MATERIALS			
ANIMATION PHOTOGRAPHY			
		TOTAL FOR CATEGORY	

Figure 1.1: Post-Production Budget Form (Cont.)

CATEGORY	UNITS	RATE	SUB-TOTAL
G. VIDEO COSTS			
FILM TO VIDEOTAPE TRANSFER	____	____	____
1" TO 3/4" WKPRT CASSETTE	____	____	____
TIME CODING	____	____	____
OFFLINE EDITING	____	____	____
ONLINE EDITING	____	____	____
COLOR CAMERA	____	____	____
BLACK-AND-WHITE CAMERA	____	____	____
DIGITAL VIDEO EFFECTS	____	____	____
TAPE-TO-TAPE TRANSFER	____	____	____
TAPE-TO-FILM TRANSFER	____	____	____
MASTER(S)	____	____	____
DUPES AND PROTECTION MASTER	____	____	____
CASSETTE DUPES	____	____	____
TAPE STOCK AND REELS	____	____	____
LABOR	____	____	____
		TOTAL FOR CATEGORY	____
H. MISCELLANEOUS COSTS			
FILM/VIDEO SUPPLIES	____	____	____
SHIPPING	____	____	____
MESSENGER(S)	____	____	____
WORKING MEALS	____	____	____
	____	____	____
	____	____	____
	____	____	____
	____	____	____
		TOTAL FOR CATEGORY	____

once the two have been synchronized. The editor uses these numbers to edit the film and sound track. Also, if the project is ultimately edited on film, these numbers allow the crew member responsible for cutting the negative to refer back to the original key numbers when conforming the final master.

Reprinting costs are any fees incurred from making additional copies of film footage that will be used more than once in a project. If the editing will be done on film, the reprinting item also covers the cost of replacing broken or torn workprint footage. This cost item does not apply to projects edited on video.

If the sound track is to be produced and edited in conventional film style, you may also have to pay for having reversal dupes made. Reversal dupes are the color or monochrome (black-and-white) copies of the edited workprint used by sound effects editors, composers and musicians during scoring sessions. Usually, producers will choose to go with monochrome reversal dupes, since monochrome copies are generally about one-third the cost of color copies. In fact, this practice is so common that black-and-white reversal dupes are often called "scoring dupes."

If the sound track will be edited electronically, you won't need scoring dupes. Instead, you'll usually transfer the edited workprint to ¾-inch videocassettes.

The dupe CRI (color reversal internegative) category covers the cost of making new negatives for the reprints needed if you conform the production on film rather than videotape.

Negative preparation/conforming and answer print preparation are two more items that apply only when you're conforming the master on film. Negative preparation and conforming covers the cost of sorting out the film rolls by key numbers, creating a key number list from the editor's edge number list and then using these keys to conform the film negative. The answer print item includes the cost of preparing a color-corrected trial print from the final release negative, to make sure that everything is in order before you duplicate the actual release prints.

You'll need to pay for a 16mm reduction answer print only if you edited the program on 35mm film but plan to distribute it as 16mm prints. This is often the case when you are preparing a program for international distribution.

The last two items listed under Laboratory Costs, 35mm protection IP (interpositive) and 35mm release print, apply only when you are distributing the finished program on 35mm film. In this case, you would usually start by making a protection interpositive print from the original negative. This is done to create a backup copy for safety reasons and for creating a duplicate negative that can be used for making the prints used in mass distribution. Finally, once you have reviewed and approved the 35mm answer print, you must pay for duplicating the 35mm release prints that you will send to the people who paid for the entire production in the first place. To save a little money, some producers use the approved answer print as one of these release prints.

Stock Shot Costs

On many projects, commercially available stock shots (crowd scenes, foreign locations, etc.) must be edited into the program during post-production. The fees for using this footage include

- library fees to cover the cost of maintaining a stock film library (if the facility has its own library)
- lab processing fees to cover the cost of preparing a print, interpositive or color reversal internegative
- rental fees for renting footage from a commercial service

Rental fees typically include a general rental charge as well as a separate assessment based on the number of feet actually used in the production. As a result, editors are often responsible for maintaining records of the stock footage edited into a production. If possible, those records should indicate the number of film frames used rather than the number of video frames, since this is how stock footage is usually billed. Many times I've had to backtrack, counting the number of video frames and then converting that number back to the equivalent number of film frames.

Post-Production Sound Costs

Sound post-production usually involves a number of steps, each with its own fees and charges. The exact number of steps depends on the complexity of the sound track and the number of enhancements (sound effects, laugh track, etc.) that the production requires.

The post-production sound costs category includes separate budget lines for each of the following:

- recording narration tracks (excluding the talent fee)
- conducting automatic dialog replacement (ADR) looping sessions (excluding the talent fee)
- locating or creating and recording sound effects
- searching and paying for stock music
- preparing a quick reference mix (scratch track)
- sweetening and mixing the sound track, including the cost of the layover, pre-lay, final mix and layback stages

- preparing the mag track, including transferring the ¼-inch production tracks, temporary mix, final mix or other sound tracks to magnetic striped film (which may then be transferred to videotape for electronic editing, if necessary)
- adding a laugh track to the final edited sound track, including equipment rental and operator fees
- making an optical sound track for film release prints (an item that does not apply to projects that will be distributed on videotape)
- purchasing any stock materials (videotape, audiotape, etc.) not covered in the other cost categories
- paying for any labor costs not covered in the other cost categories

For more information on the sound post-production process, see Chapter 6.

Scoring Session Costs

Sound scoring sessions can encompass a variety of fairly specific activities. As a result, if your production requires a separate session to prepare and record a musical score, you should probably treat it as a separate budget item. Expenses encountered during a typical scoring session include

- purchasing the rights and licenses to any copyrighted music to be used in the score
- paying the composer and lyricist, if any, for any original music that will be used in the score
- hiring an arranger and paying music copyists and printers to write the scores for each musician
- hiring one or more directors to manage and conduct the orchestra
- hiring musicians
- renting instruments and paying for instruments to be hauled in and out of the scoring stage or studio
- renting a video or film projector if the composer, arranger or conductors want to time their work to a "live" playback of the program
- hiring a projectionist
- renting the scoring stage or studio
- paying for the scoring stage crew
- paying for the mixdown of the recorded score

If you have arranged a package deal with a scoring facility, be sure to ask which of the items listed above are included in the package price. Also, remember that union rules require a minimum three-hour "call fee" for each musician and that you must give musicians at least 96 hours' notice if the scoring session has been canceled or postponed.

Special Facilities Costs

The special facilities category includes the cost of renting a film or video screening room, plus any special facilities fees that are not covered in other categories. For theatrical film projects, you will normally need a film screening room for reviewing dailies, as well as for viewing and evaluating the conformed workprint before giving the go-ahead for final conformation.

For video productions, you will need a video screening room for reviewing edited workprint copies before the project is approved for online editing. Although some producers prefer a large screening room with a projection TV system, I've found that video screenings generally tend to be more effective when you are watching the program in the same setting in which the finished program will be seen (e.g., a home living room setting, a corporate seminar room, etc.).

The last two items in this category—equipment rental and labor—cover any equipment or labor costs that are not included in the rental price for the screening facility.

Opticals and Animation Costs

The opticals and animation category includes the costs associated with preparing the graphics and art needed for post-production. The various budget items in this category are described below:

- Artwork covers the price of preparing full-color drawings, title cards and any other illustrations that will be added to the program during post-production.
- Title photography includes the cost of photographing the titles on film. This item applies only to projects that will be conformed on film.
- Matte photography refers to the costs involved with photographing "HiCon" or "traveling" mattes (opaque images that animate to match the movements of an image for matting purposes).
- HiCon developing and printing covers the lab fees for processing any matte photography.
- Optical effects includes the cost of preparing and shooting special effects (fades, dissolves, wipes, etc.). This item usually applies only to projects that are conformed on film, since the cost of effects for programs finished on video is generally included in the online editing fee.
- Lab processing fees apply only to productions conformed on film. They include the cost of processing all film opticals.
- Titles amortization is an important item for producers of TV series. Since each episode of the series usually begins with the same main title sequence, producers can amortize the cost of creating that sequence over the length of the series.

- Animator and materials fees cover all of the costs associated with drawing and generating any animation used in post-production, including any computer animation.
- Animation photography includes the fees for photographing the animation on film or recording computer-generated animation on videotape.

Because art needs differ from one production to the next, the total budget for this category will vary considerably from project to project.

Videotape Costs

This category covers most of the costs related to the preparation, coding and editing of the videotapes used in electronic post-production. Those costs include any fees for transferring film to tape (using a double or single transfer system, with or without color correction), making copies of "window dupes" (videocassettes that display time code in a small window on the screen) or adding any other time code during or after the transfer process.

Videotape costs also include all offline and online editing charges. Usually, the editing prices quoted by post-production facilities cover the cost of the editor and assistant editor, plus the rental fee for using the editing room and equipment. When these labor costs are not part of the facility fee, you should list them as a separate budget item.

Some post-production facilities also include any color camera, monochrome camera and special effects equipment costs in their base fees. Check to make sure that this is the case when you are comparing the prices quoted by different facilities.

Other fees in the videotape costs category include charges for

- tape-to-tape transfer, particularly the costs of "bumping up" the production (transferring from one videotape format to another)
- tape-to-film transfer, including the costs of transferring the videotape sound track to the ¼-inch magnetic film track, preparing a "35/32 A-wind" optical track and processing and printing the final delivery print
- tape stock, tape reels and other items of this type not covered in the previous categories
- labor costs not covered in the facility fee

As I've mentioned, the items included in a post-production studio's base rate vary from facility to facility. Be sure to find out what the base fee covers before committing a project to a particular post-production studio. Otherwise, you may be in for an unpleasant surprise when you open the facility's final bill.

Miscellaneous Costs

This category serves as a catchall for all those unpredictable, but inevitable, costs that pop up during the course of any post-production session. Those costs include the price of video and film supplies (splicing tape, film leader, etc.) not listed in the other categories, shipping and messenger fees, and the tabs for any meals devoured during midnight editing sessions. Needless to say, these little charges can mount up. In fact, one video producer I know adds up the subtotals of the other budgetary categories and then figures in 10% of that sum to cover miscellaneous costs.

CONCLUSION

In this chapter, I began by discussing the many creative, technical, scheduling and budgetary factors that producers and filmmakers must keep in mind as they consider the various post-production alternatives. As I pointed out, the electronic post-production alternative makes the most sense when the program will be distributed on videotape and when post-production sound requirements aren't too complex. However, recent technological developments promise to make electronic posting a very viable option on an even wider variety of projects. For more information on these developments, see Chapter 5.

For many producers, cost is the ultimate consideration. Given the right conditions, electronic posting can result in considerable cost savings. Some of those savings are illustrated in the comparison budget offered earlier in the chapter, and in my discussion of the detailed post-production budget that ends the chapter.

Electronic posting can also save time, and time saved usually means money saved. Even more important, if air dates or delivery schedules are on the line, time saved can also mean careers saved.

After weighing all the variables, producers and filmmakers must decide for themselves whether some form of electronic post-production is the right post-production process for their particular project. To help them understand the steps involved in that process, I've analyzed several case studies in Chapter 2. As those case studies show, electronic posting can involve a number of different steps, depending on the conditions and circumstances surrounding the production.

NOTE

1. *Society of Motion Picture Technicians and Engineers Journal* (June 1982): 552.

2 The Electronic Post-Production Process

As I mentioned at the end of Chapter 1, the specific steps involved in electronic post-production differ from project to project, depending on the scope and complexity of the production. With this in mind, I'll be using four different types of programs to illustrate the electronic post-production process:

- a weekly, one-hour television drama
- a feature film produced for theatrical distribution
- a music video
- a television commercial

Taken together, these case studies should provide film and video professionals with a fairly complete picture of the electronic posting process.

A WEEKLY TELEVISION DRAMA

Figure 2.1 shows the main steps involved in the post-production of a typical television series. As our example, I've selected "Fame," a weekly musical drama that is currently being produced for syndication following a run on network television. "Fame" is a relatively elaborate production: each one-hour program may incorporate single and multiple camera work, location and sound stage shooting sessions, and prerecorded track playback for the musical scenes. The episodes are shot on film, and the film is then transferred to videotape for electronic post-production.

Production and Film Processing

"Fame" is shot on 35mm film using conventional lighting techniques and #5247 film stock. According to Frank Merwald, the associate producer of "Fame,"

Figure 2.1: Post-Production Flow Chart for a Television Series

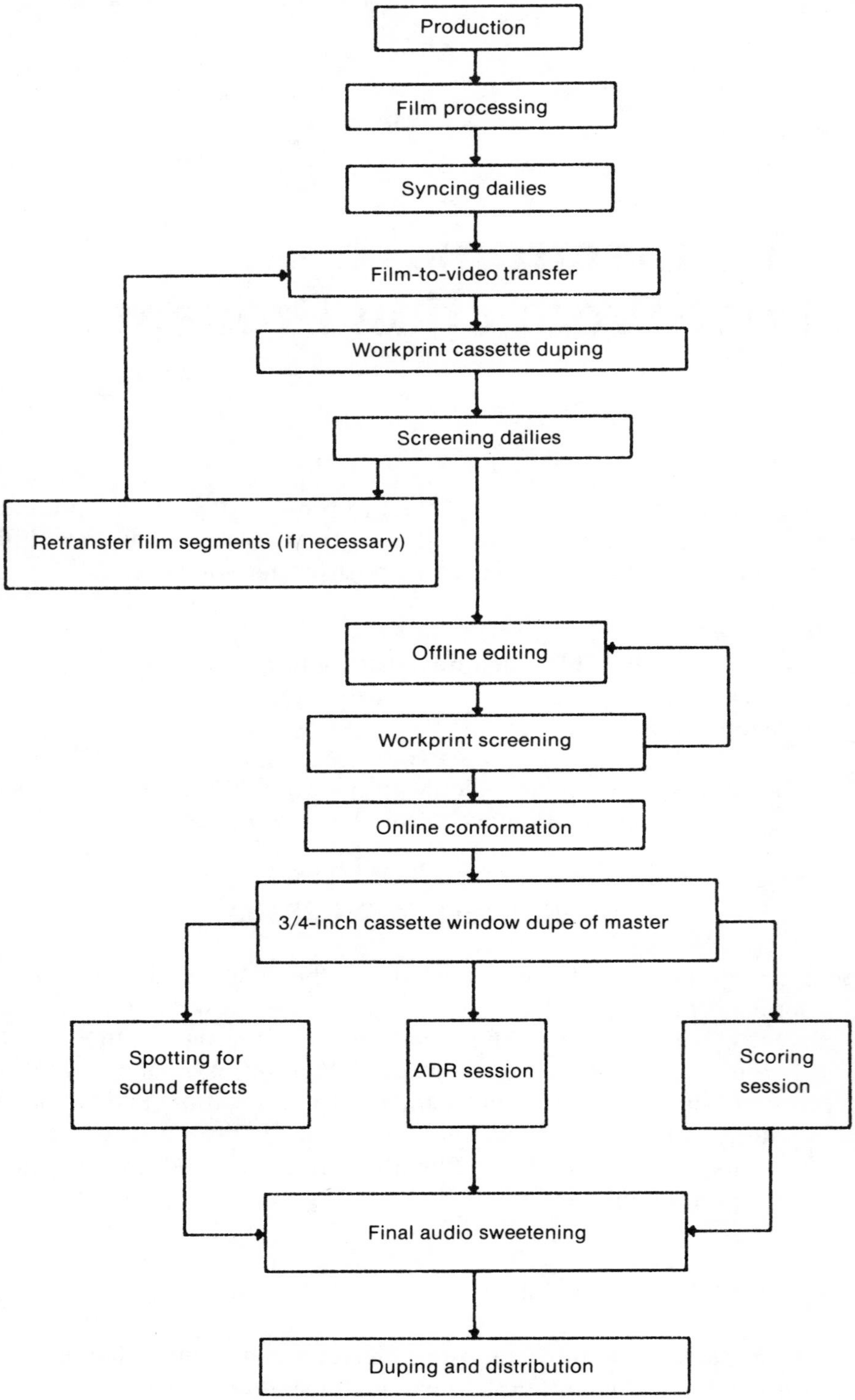

Production

Film processing

Syncing dailies

Film-to-video transfer

Workprint cassette duping

Screening dailies

Retransfer film segments (if necessary)

Offline editing

Workprint screening

Online conformation

3/4-inch cassette window dupe of master

Spotting for sound effects

ADR session

Scoring session

Final audio sweetening

Duping and distribution

no special adjustments are necessary during shooting to accommodate electronic post-production.

Each episode requires seven days of shooting, with the typical production day lasting 12 hours. After each day of shooting, selected takes are delivered to the MGM film processing laboratory for overnight development. At 6:00 a.m. the following morning, the developed "dailies" are transported to Randken Corp.'s facilities for sound synchronization, as described below.

Synchronizing the Sound to the Dailies

The original post-production process for "Fame" required Randken to transfer the ¼-inch nagra sound track from the day's shooting to 35mm magnetic (mag) film overnight so that crew members could synchronize and mark the sound and picture.

This syncing process was completed by 9:00 a.m., at which time Randken delivered both the negative and the mag track to the video post-production facility. By around 10:00 a.m., the video facility would begin transferring the film to videotape using the Rank Cintel telecine system.

This process changed in 1985, when "Fame" began using a proprietary system developed by Randken Corp. and Complete Post, Inc., to transfer the ¼-inch nagra track directly to precoded 1-inch videotape. As a result, "Fame" no longer uses magnetic film for transferring the sound track.

The Film-to-Tape Transfer Process

Before the film-to-tape transfer process can start, the film negative must be loaded on the Rank Cintel telecine system and the mag sound track (if the mag track process is being used) must be placed on the mag sound machine. This is known as a "double system transfer."

Occasionally, the mag track may not be ready at transfer time, usually because of some problem in the laboratory processing or mixing stages. If this is the case, the post-production crew can still transfer the negative without sound (known as "MOS") at the scheduled time. Then, when the mag track is ready, it can be dubbed onto the transfer tape and synchronized with the video.

Once the machines are ready, the associate producer and telecine operator (also called a video colorist) view the first couple of takes on the reel, to check and compare color quality. Any color adjustments, if necessary, should be made before the actual transfer process begins.

To maintain a high degree of color consistency, the crew member in charge of post-production will usually use the same telecine system, the same video colorist, the same color viewing monitor, and the same videotape recorder (VTR) for all of the transfers on a given episode. If this is not possible, an alternate method for maintaining color quality is to store sample frames from different takes on a frame-store device. Then, the video colorist can adjust the color from new takes to match the stored frames. For more information on making these kinds of color adjustments, see Chapter 4.

After the video colorist enters the color corrections, the machines roll, and the film negative and mag track are transferred to 1-inch videotape. To accommodate electronic editing, the videotape used in the transfer is "precoded" with SMPTE time code.

To help save time in editing, "Fame" uses a separate reel of precoded 1-inch videotape for each transfer session. Because there are seven transfer sessions (one for each day of shooting), "Fame" ends up with seven transfer tapes. The transfer reel from the first day of shooting is labeled "Reel 1," with the time code starting at hour 01, the reel from the second day is labeled "Reel 2," with the time code starting at hour 02, and so on.

One exception to this rule is the transfer of footage from multiple-camera shooting sessions, which for "Fame" averages between 5000 and 7000 feet of film (55 to 78 minutes of video) for each episode. For multiple camera shoots, the footage from each camera is transferred to a separate reel.

Transferring and maintaining material on separate reels offers two important benefits. First, it shortens editing time by eliminating the long search and cue periods necessary when all of the material is stored on one or two reels of transfer tape. Second, it gives the editors more freedom and flexibility in creating dissolves and other visual effects between shots.

When all of the transfers are completed, each 1-inch reel of videotape contains the color-corrected video from a day's shooting (or the video from one camera in a multiple-camera shoot), one or two audio tracks and longitudinal time code. In other words, the reels are ready for editing—as soon as the producer and director have a chance to review and evaluate workprint copies.

Making Workprint Copies

Once the transfer process is complete, crew members make two ¾-inch videocassette copies of the 1-inch transfer reels. These "editorial workprints" are exact copies of the 1-inch tapes, except that the SMPTE time code numbers will be imprinted on the picture in one or both copies. Because the time code numbers appear in a small window on the screen, these workprint copies are also called "window dupes."

Screening the Workprint Dailies

Around 1:00 p.m. the producer, director and director of photography review the ¾-inch workprints at an MGM screening room. They select shots and scenes that will be included in the edited program, noting the time code numbers as reference points for the editors. They also note any problems with dirty film negatives, color imbalances, and sound synchronization that will need to be corrected in the next regularly scheduled transfer session. As a last resort, problem scenes can also be redone during the next day's shooting.

Correcting Special Problems

As they screen the workprints, the producer and director may note problems that require a new negative transfer. Usually, these problems are corrected with a frame-accurate telecine/VTR control system such as the automatic video replacement system (AVRS) or the time logic controller (TLC) described in Chapter 4. Using this type of system, a dirty negative discovered during the screening can be cleaned and retransferred onto the same segment of the original 1-inch transfer tape. This eliminates the need for resyncing the sound and time code on the transfer tape—a time-consuming process.

A telecine/VTR control system also allows the video colorist to make color corrections to only the problem parts of the tape, saving the post-production team from having to retransfer the entire reel.

After the post-production crew makes all of the required fixes on the 1-inch master tape, they dub a new "pickup" cassette for use in the preview and editing stages.

Editing the Workprint: The Offline Editing Stage

The offline editing stage is the stage of post-production in which editors prepare a videotape workprint of the entire episode, along with a computerized edit decision list (EDL) that they will use in editing the finished master program tape. The process of assembling that final master tape is called "online editing" or "video conformation." For more information about offline and online editing, see Chapters 5 and 6.

For the typical episode of "Fame," the offline editing period begins on the second day of shooting, runs for 15 days and incorporates three distinct stages. In the first stage, also called the first cut or director's cut stage, the editor assembles the shots and takes selected by the production personnel during the workprint screening sessions. At this point, the focus is on establishing pacing and flow, with little regard for total running time. In the second stage, the editor shows this first-cut tape to the producer and executive producer, who indicate where they would like changes made to improve the pace, flow and timing of the program. Finally, in the third stage of offline editing, the editors take the workprint corrected in stage two and make any last minute changes. They also put the finishing touches on the workprint, making sure that they have made the proper adjustments and allowances for the main title sequence, commercial breaks and closing credits.

During the offline editing process, the editors work with ¾-inch window dupes dubbed earlier from the 1-inch transfer tapes. Taking the selected shots and scenes from these ¾-inch cassettes, they assemble the final workprint on a conventional videocassette editing system. In the case of "Fame," the final workprint is pieced together on a Grass Valley Group, Inc. editing system employing Sony BVU-800 videocassette recorders.

At each stage of the offline editing process, the system generates a computerized EDL. By the end of stage three, then, the editor has three separate decision lists. These lists are "cleaned" and "traced" (see Appendix B) and the resulting final EDL is then used in the online editing process described below.

A final word about the sound track. During each stage of editing, the dialog tracks are split alternately between tracks one and two of the edited workprint cassette. This same procedure is maintained in online editing, so the mixer is able to equalize the tracks during the audio sweetening process.

Video Conformation: The Online Editing Stage

Video conformation, or "online editing," is the stage of electronic post-production in which scenes and segments from the original 1-inch transfer tapes are pieced together to form the final edited master. In conventional film editing terms, this would be analogous to cutting the negative, performing all opticals, completing basic sound editing, correcting color timing and creating the final release prints—all in one stage.

As described in Chapter 6, online editing is an automated, computerized procedure. First, the 1-inch transfer reels containing the original production material are loaded onto playback VTRs. Then, the segments selected during offline editing are copied, in the correct sequence, to a new 1-inch videotape called the edit master reel. In the middle, to cue up the tapes and control the whole process, is an editing computer. And controlling the computer is the edit decision list—the "program" generated in offline editing that tells the computer which segments appear in which sequence and what types of transitions (dissolves, straight cuts, etc.) should join the different segments.

Online editing for an episode of "Fame" takes place over two workdays. On the first day, the 600 to 700 edits that go into each episode are assembled onto the master reel over a 9-to-10–hour period. This process creates a textless version of the program that serves as the foreign version (without titles) and as the basis for the domestic version (with titles).

On the second day of online editing, the main titles, opening "teaser" sequence and closing credits are added to a copy of the textless version. This tape then becomes the edited master reel for the domestic version of "Fame."

For the typical episode of "Fame," the entire online process adds up to 14 to 16 hours of billable editing time.

Sound Effects, Looping and Scoring

Once the video conformation process is complete, the post-production team delivers a ¾-inch window dupe copy of the edited master to the sound effects editor for spotting—the process of reviewing the tape for locations where any special sound effects are necessary. The editor also checks the tape for any scenes that might require automatic dialog replacement (ADR) work.

The sound effects editor and audio technicians record and synchronize all sound effects and "looped" dialog in MGM's sync room. The actors who need to loop lines of dialog also visit the sync room, where they watch the program on a large-screen video projector and time their lines accordingly.

A large-screen video projector is also used to score the many musical productions. The conductor and musicians view the window dupe version of the program as they play, and the music is recorded onto a 24-track audiotape.

Audio Sweetening

As discussed in Chapter 6, audio sweetening is the stage of post-production in which the sound mixers equalize the sound tracks, add sound effects and perform any music mixdowns.

With its many musical productions, "Fame" schedules a relatively long sweetening period—almost four full days per episode. On the first day of sweetening, members of the sound team pre-lay the music. On the second day, they pre-lay the sound effects and any ADR lines and then equalize (pre-dub) the split production tracks. On the third day, they complete the final mixdown of the tracks that were pre-layed on the previous two days. After finishing the mixdown, the team dubs a ¾-inch cassette copy of the program for the producers to review and evaluate.

That evaluation takes place on the morning of the fourth day. If the producers suggest some changes, those adjustments are worked into the final mixdown that afternoon. Once the changes are made, the final sound tracks are layed back (recorded) onto the 1-inch edited master videotape.

Because "Fame" is distributed both domestically and internationally, the sound team actually goes through two layback sessions. For the domestic version, they layback mixed sound tracks onto both tracks of an edited master. For the foreign version, they separate the tracks, recording the dialog on track one and the mixed music and effects tracks on track two. This allows foreign countries to dub the dialog in their own language.

Distribution

For domestic distribution, each station that has purchased "Fame" receives one copy of each episode per week. About half the participating stations receive their copies through a satellite feed, and the other half are shipped videotape copies. The version of the program sent to the stations includes the program video, the titles and a mixed sound track.

The international distributor receives two versions of the program: one with text and titles and one without. As described in the previous section, the textless version contains split sound tracks to accommodate dubbing. Both versions are converted to the European phase alternating line (PAL) broadcasting standard, with the text version going to English–speaking countries and the textless version going to non-English–speaking countries.

A FEATURE FILM

As I mentioned in Chapter 1, the producers of feature films have historically resisted electronic post-production—with some justification. After all, until very recently, it has been difficult to generate an accurate negative cutting list through electronic post-production. And generating that cutting list is a key step in the film editing process since, in almost all cases, features produced for theatrical distribution are delivered to theaters as 35mm prints prepared from a negative that has been cut and spliced in the conventional film manner.

Other problems that must be overcome in the electronic post-production of feature films include the following:

- the limitations of conventional electronic sound posting technology that I discussed in Chapter 1
- the concern that screening a videotape workprint on a TV monitor does not give producers and directors an accurate "feel" for how the film will look to an audience in movie theaters
- the reluctance of many film editors to make the switch to new and unfamiliar editing equipment

The first two problems—generating an accurate negative list and working with film sound—are being solved through advances in technology. (See the rest of this chapter, plus the sections on the Montage, EditDroid and SoundDroid systems in Chapter 5 and the PAP system in Chapter 6.) The last two problems will only be solved with time and experience, as more film professionals grow accustomed to working with electronic editing technology.

One feature film project that successfully made the switch to one form of electronic post-production is *Oh God, You Devil,* a Warner Brothers production starring George Burns, directed by Paul Bogart and edited by Randy Roberts and Andy Zall. *Oh God, You Devil* was produced on 35mm film, and the 35mm dailies were transferred to ¾-inch videotape. The videotape was then edited on a conventional computerized editing system, yielding a finished videotape workprint plus a time code–based EDL. The EDL was then converted to a negative cutting list, which the editing team used to conform the film negative. Figure 2.2 shows the steps involved in producing and editing a feature film.

The Production

Oh God, You Devil was conceived, written and filmed as a conventional theatrical feature. No special adjustments were made during filming to accommodate electronic post-production, with one exception. Although the director and producer screened dailies and compiled lists of editing notes in the usual manner, no editing was done during filming, as would typically be the case. Instead, all editing occurred after all the dailies were transferred to videotape.

Processing and Syncing the Dailies

The film negatives from each day of shooting were processed in the normal manner, and the ¼-inch nagra sound tracks were transferred to magnetic film. Then, the negatives and magnetic sound track were synchronized, the edges of the negatives were coded with frame numbers and the dailies selected for screening were printed.

Figure 2.2: Post-Production Flow Chart for a Feature Film

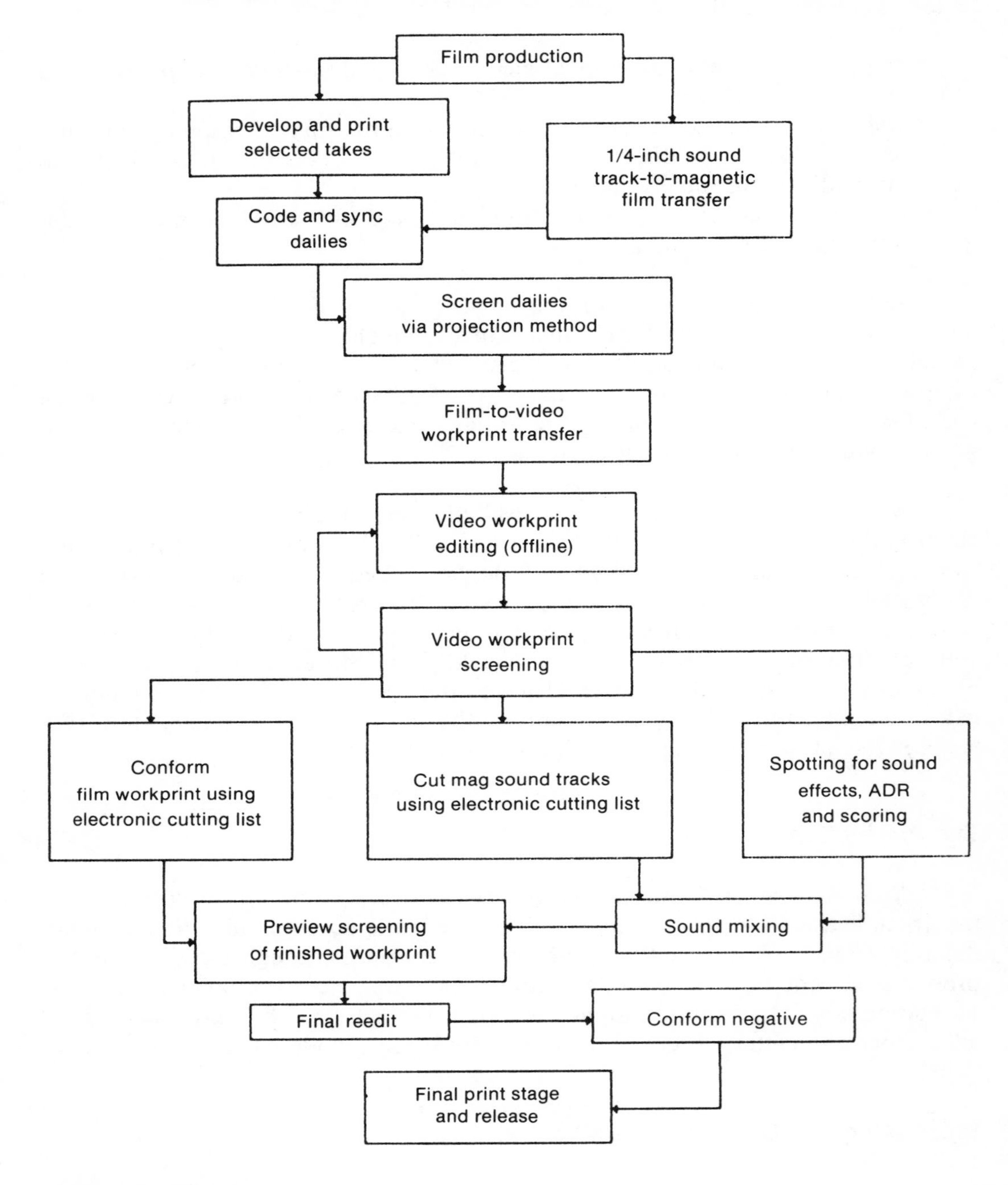

Screening the Dailies

The dailies were reviewed in a film screening room, using the normal film projection process. Because the production was shot in cinemascope, using the conventional projection process allowed the director to see how different shots and scenes would look when the film was shown in movie theaters and to check for scene densities and shot composition.

Film-to-Tape Transfer

Once they had been screened, the dailies were loaded onto a Rank Cintel system for transfer to ¾-inch videocassettes, with one videocassette used for each 1000-foot reel of film. During the transfer process, scene, take, edge and time code numbers were burned into the video picture. Since the film was shot in cinemascope, there was plenty of room at the top and bottom of the video image where this information could reside unobtrusively.

The different numbers burned into the image serve a variety of purposes. For example, along with helping the post-production crew hand log edit points, the scene and take numbers provide advertising and promotional staff with a handy reference for selecting the footage to use in their respective campaigns. The time code numbers also serve as a reference during offline editing, and the edge code numbers allow the post-production team to double-check the cutting list that will be generated following offline editing.

Offline Editing

On feature film projects, offline editing is the stage of post-production in which the decision to go electronic begins to pay off. For *Oh God, You Devil,* offline editing of the video workprints took place over a five-week period and yielded three different edited versions of the production. All of the editing was performed on a computerized video editing system that included a video switcher and audio mixing board.

The video switcher more than pays for itself, since it allows the director and producers to view trial opticals and special effects on videocassette before they commit to expensive film opticals. For *Oh God, You Devil,* the offline editing team generated temporary titles and keyed them into the picture, performed dissolves and fades between shots and even created some makeshift animation. In fact, because they were able to screen the edited workprint that included all of these "video opticals," the film optical house completed its work correctly on the first try.

During offline editing, a sufficiently accurate sound track was prepared by recording two overlapping sound tracks on the edited workprint cassette and adding temporary music. As described in the next section, the actual sound track was prepared in a separate sound mixing session.

Once the offline team finished editing the video workprint, Complete Post, Inc. generated an electronic negative cutting list by correlating the time code numbers on the edited workprint with the film edge code. Using the electronic cutting list, the feature editor conformed (edited) the 35mm dailies over a five-day period, ending up with a film workprint that matched the edited video workprint. Essentially, then, the entire workprint editing process required six work weeks: five weeks for offline editing and one week for conforming the dailies.

Sound Mixing

Sound editing and mixing, including film scoring, ADR, and all sound effects work, was completed at a Warner Brothers' sound facility. The sound work was performed in the conventional film manner, except that editors were able to use the electronic cutting list as a guide for cutting and splicing the film mag track.

Because *Oh God, You Devil* was already sold for TV distribution, the sound editors prepared both a TV track version (sound on *one* track, since this predated the era of stereo TV) and a three-track theatrical version.

Final Editing

Only minimal "cleanup" editing was required after the film workprint was prepared following the offline session. For the most part, this final editing involved minor scene lifts requested by the director after preview screenings and some take changes (primarily to modify objectionable language) in the TV version.

A MUSIC VIDEO

Music videos fall into two broad categories: concert videos and concept videos. In concert videos, the musicians are filmed performing a song before a live audience, usually in a concert hall or arena setting. In concept videos, the song plays in the background as the musicians and other performers interpret it through a series of dramatic, and often quite abstract, vignettes. Recently, however, many videos have mixed the two forms, combining concert footage with dramatic sequences.

In this section, rather than follow the production path of a single music video, I'll speak generally about the steps involved in producing and "posting" the various types of videos. Figure 2.3 is a flow chart that illustrates the different stages of a typical music video production.

The Production and Film Processing

Most music videos are shot on 35mm or 16mm film, using a single camera. Multiple-camera shooting, when done, is usually reserved for filming the live performance portions of the production.

Figure 2.3: Post-Production Flow Chart for a Music Video

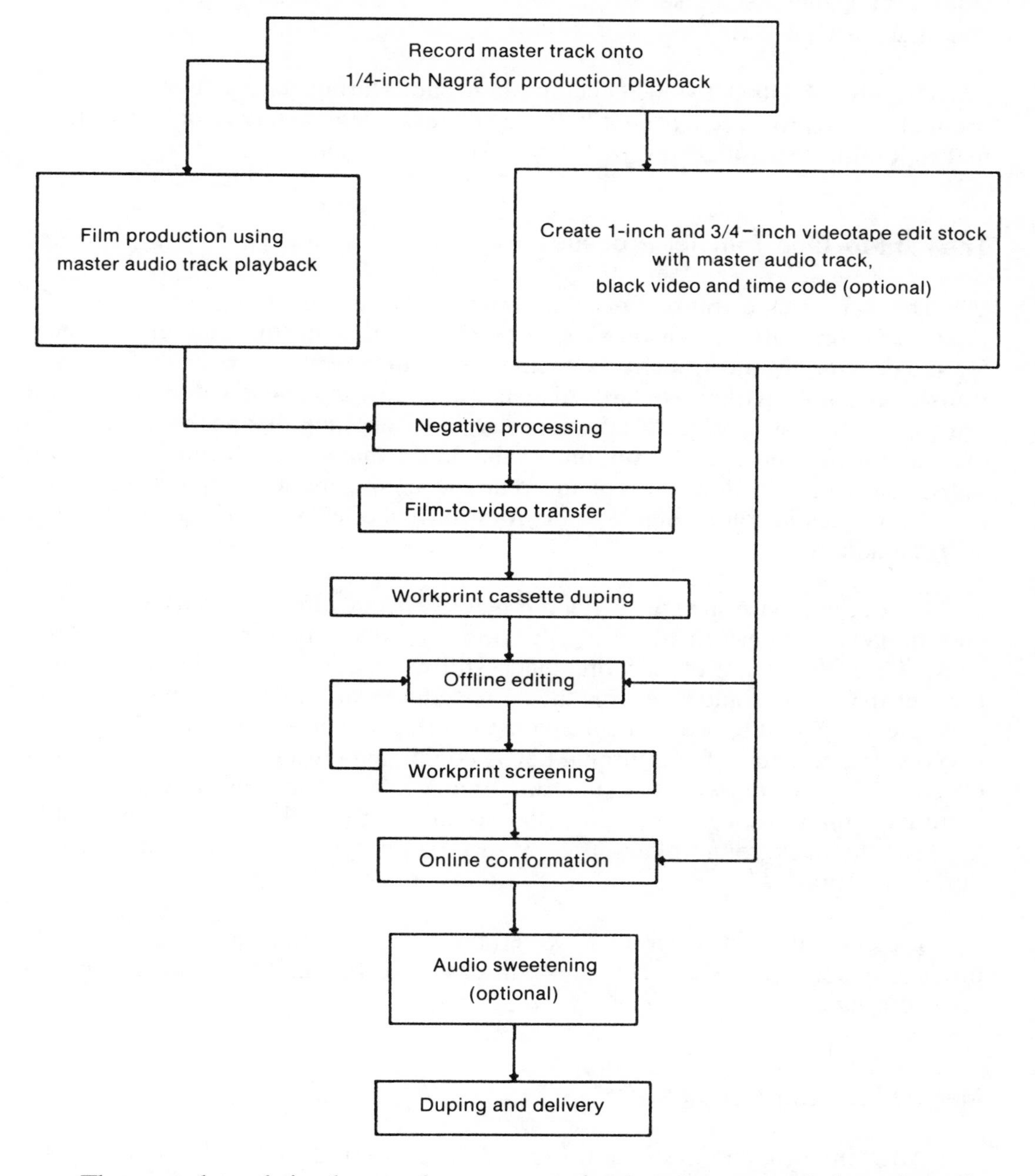

The sound track is almost always a premixed and prerecorded version of a single song, a master copy of which the record company will usually provide. The record company's master is then transferred on a ¼-inch nagra recorder (with 60 Hz crystal sync pulses), for on-location playback.

During this transfer process, some music video producers, including Simon Straker of Pendulum Productions, will also record the master sound track and a time-coded black video track onto ¾-inch videocassette and 1-inch videotape stock. Later, they use these "pre-blacked" tapes as the stock for their workprint and final master tapes.

Usually, all 4000 to 5000 feet of film shot during a typical music video production are processed, primarily because the conceptual nature of many videos makes it difficult to single out preferred takes on location.

The Film-to-Tape Transfer Process

The film-to-tape transfer process for music videos is similar to the transfer process for dramatic television series described earlier in this chapter. However, for music videos, the transfer is a single-system process, since the footage is transferred MOS (without sound). Also, in processing the music video footage, the transfer crew may decide to add some stylized touches or special effects. For example, using some of the options available on the Rank telecine system, the video colorist might slow down or speed up the action, create or reposition zooms, adjust for realistic color, change colors for a surreal effect or enhance the image to sharpen details.

In addition, on some music video projects, special effects such as matting and chroma keys are performed during the transfer stage. For various technical reasons, these effects are often more successful when executed during film-to-tape transfer than they would be during the online editing stage. For example, the rough edge "tearing" that appears in chroma keys performed during online video editing is less likely to occur if the chroma key is completed during the transfer process. Of course, performing effects such as chroma keys during film-to-tape transfer requires some advance planning, including knowing exactly which scene will be used for the background plate in a key so it can be played against a blue screen during the transfer.

Typically, the film for a music video production is transferred to tape in the order in which it was shot. This makes it easier for the director to locate specific takes during editing.

Making Workprint Copies

After the original production footage is transferred to 1-inch videotape, the post-production crew makes a ¾-inch window dupe of each 1-inch tape for use in workprint editing. The time code on the window dupes must match the time code on the 1-inch masters. In addition, if the offline editing is to be performed on a computerized, time code–based system (which I strongly advise), the window dupes must also include the longitudinal time code described in Chapter 3.

To save time in the offline editing stage, I have occasionally used two identical window dupes of any transfer reels that contain more than 45 minutes of program material. Later, during offline editing, one window dupe can be cued to a point near the front of the reel, and the other can be cued to a point near the end of the reel. This will cut down on the long search periods necessary when the editor must scan the full length of a cassette for a particular scene.

Editing the Workprint: The Offline Editing Stage

During offline editing, the director and editor view the workprint cassettes together, selecting takes to be included in the edited video and jotting down the appropriate edit points. Because the workprint cassettes duplicated from the transfer reels do not contain sound, the director and editor must often play the master sound track in sync as they preview the tapes, to make sure that there are no problems with lip syncing or the timing of the performance. In some cases, the editor will actually record the sound track in sync on the workprint tapes before the director arrives for the screening.

The typical music video requires between 24 and 30 hours of offline time to create an edited master workprint. Once the post-production team completes the edited workprint, they are ready to generate the edit decision list—either by hand logging the edit points or by using a computerized editing system to produce the list on punch tape or a floppy disk. Any special effects that will be added during the online conformation stage must be noted and hand logged. To save time and money during online editing, the special effects log should include the exact time code for the locations where each effect will begin and end.

Video Conformation: The Online Editing Stage

As explained earlier in the chapter, video conformation is the expensive process of using a computerized editing system to create the final edited master from the original 1-inch transfer tapes. At this stage of post-production, the accuracy of the EDL generated during offline editing becomes critical. Generally, an inaccurate editing list of the type obtained from many low-end editing systems will increase online editing time by about half. With online editing fees running to several hundred dollars an hour, it's not hard to see that money saved by purchasing or renting an inadequate offline editing system will quickly turn into money wasted during online editing.

One trick that I find helpful is to record the edited workprint master generated during offline editing onto the 1-inch tape that you will use for assembling the final program during online conformation. Once the computerized editing system starts the automatic assembly process, it will replace the workprint video shot by shot, allowing you to check the accuracy of each online edit against its offline predecessor as the edits are being performed.

I've also found that the best approach for adding special effects is to complete the automatic online assembly for the entire program and then go back to edit in the effects, rather than to disrupt the process to add the special effects during auto assembly (except for dissolves and wipes). That way, you don't tie up expensive special effects equipment for the entire online period—only the time required to add the special effects at the end of the session. You also have the advantage of viewing clean, finished video on both sides of each special effect that you do edit in.

Audio Sweetening

Because the sound track for most music videos is a fully finished master track supplied by the record company, most music video productions are able to skip the sound sweetening stage. However, conceptual videos sometimes include sound effects, an announcer track or audience reaction sounds that are added during an audio sweetening session. For more information on audio sweetening and "sound posting," see Chapter 6.

Distribution

At this writing, the typical distribution deal for a music video requires the producer to supply the record company that is paying for the video with two 1-inch copies of the master (the original and one protection dupe) and two ¾-inch cassette copies, along with all of the original production film and 1-inch transfer reels. Usually, the record company will order any additional copies that they need through the original production company.

A TELEVISION COMMERCIAL

The post-production of television commercial spots can differ drastically from one project to the next, depending on the complexity of the production, the available budget and the preferences of the commercial house that is producing the spot. For example, a 30-second commercial that consists of three or four dialog scenes and a shot of the product will require much less post-production work than a 60-second spot that includes multi-image electronic effects and an elaborate music track.

Figure 2.4 shows the steps involved in producing and editing a typical TV commercial. As this flow chart suggests, most commercial houses still prefer to stay with conventional film techniques through the initial production stages,

Figure 2.4: Post-Production Flow Chart for a Television Commercial

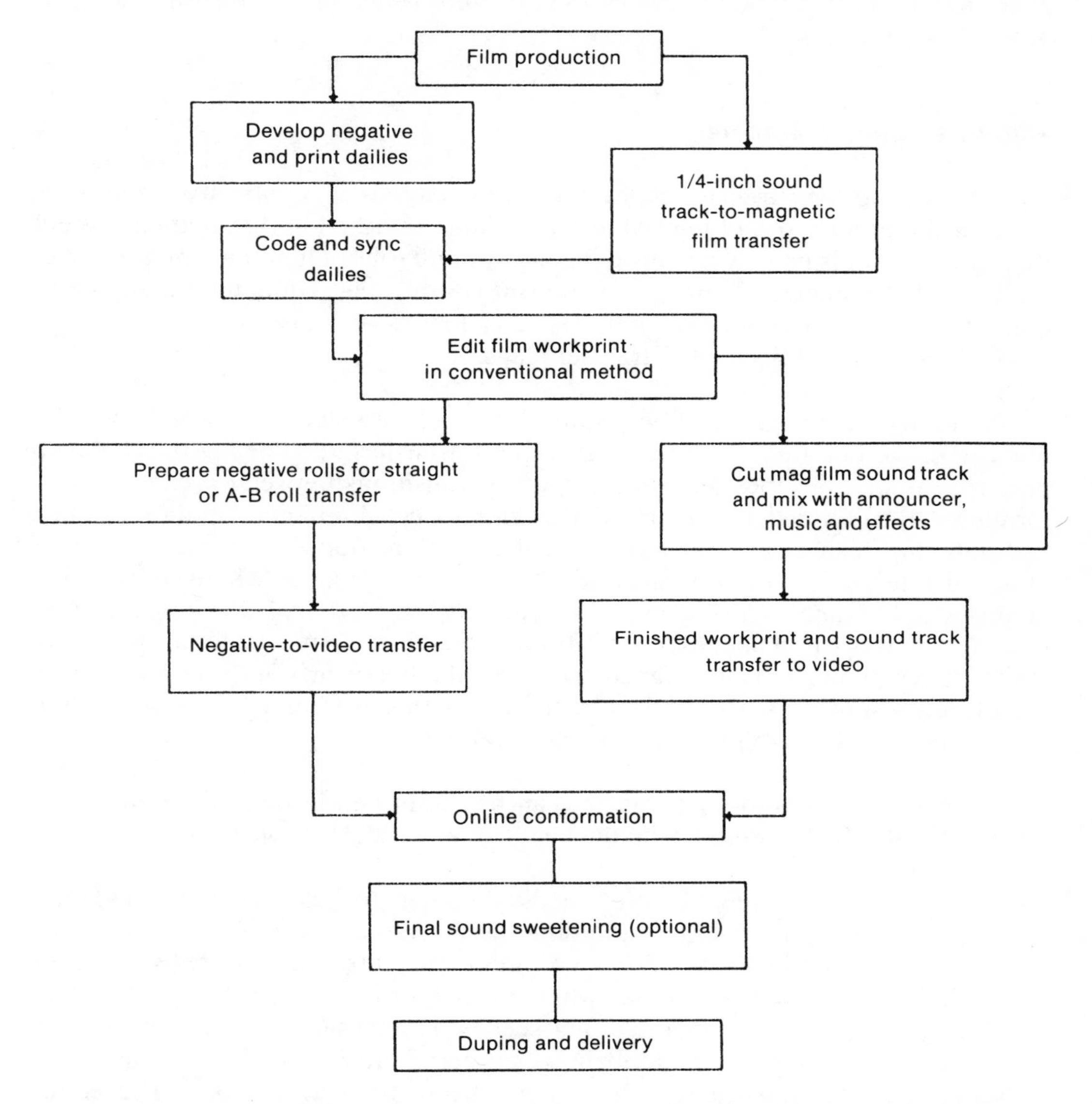

striking dailies and editing the workprint in the traditional film manner. By the end of this process, the production team will have in hand an edited film workprint, completed mag sound track, videotape transfers of the original film footage and a frame-count list. They then take these materials to a video editing facility, where they conform the production electronically.

By conforming the production on video, the commercial house saves considerable time and expense. It is also able to take advantage of the many features that computerized editing systems offer, particularly in the areas of electronic opticals, graphics and digital special effects.

On the pages that follow, I'll focus on two important stages of commercial post-production that tend to present the most problems: film-to-tape transfer and online conformation.

Film-to-Videotape Transfer

The film-to-tape transfer process for commercials is similar to the process used for the other types of television production described in this chapter, except that the negative is usually arranged on one or more rolls of film that correspond to the count list prepared during the workprint editing stage. In addition, any color correction, noise reduction or variable speed (slow motion, fast motion, etc.) work is usually performed during the transfer stage.

Complex commercials often require "A-B roll transfers," in which two different reels of film (an A roll and a B roll) are transferred, allowing the editor to perform dissolves, wipes and other types of transitions between scenes. In performing A-B roll transfers, the transfer team must be sure to overlap the incoming and outgoing scenes, to compensate for the different frames-per-second rates of film and television. Film operates at 24 frames per second, whereas television operates at 30 frames per second (see Chapter 3). An overlap of six to eight frames in each A-B roll will make up for this frame difference, compensate for instability at film splice points and allow for slight cut point adjustments, if necessary, during the subsequent mixing process. Failure to include this overlap will usually result in black frames and instability at every crossover point.

A-B rolls for film-to-tape transfer can be performed in three different ways, depending on which method suits the needs of a particular project:

- Transfer the entire amount of unedited footage (or just selected takes) onto two rolls of videotape simultaneously (called "double record"). This method is well suited to relatively simple projects with tight delivery deadlines, since there is no delay while waiting for film lab work to be finished.
- Record spliced but unedited film scenes on alternate videotape rolls. This method will sometimes apply in situations where film workprints have been finished but no negative cutting has been, or will be, done. The editor simply makes a foot/frame count list, which the video editor then uses for converting film footage length to television time. For conversion tables, see Chapter 3.
- Transfer correctly spliced and edited (with the appropriate overlaps) A-B rolls, synchronized with a complete audio track. These A-B transfer rolls are then synchronized and mixed together during video editing, using a conventional method of converting film footage to television time. In fact, some computerized video editing systems feature functions that automatically convert foot/frame numbers to the correct time code references.

A few final notes on A-B roll preparation. First, make sure that the footage needed for transitions (dissolves, wipes, etc.) includes the necessary eight-frame overlap counted from the *last* transition frame of the outgoing scene and the first frame of the incoming scene. Second, keep in mind that film editors usually use the midpoint of a transition for duration timing purposes, whereas most videotape editors count from the first frame of the transition.

Finally, if the film cutting list must be converted to TV time, I recommend using SMPTE non-drop-frame time code. That way, the mathematics involved in the conversion is fairly straightforward and any actual time-length discrepancy is negligible. For more information, see the discussion of film-to-TV time conversion in Chapter 3.

Online Video Conformation

Usually, video editors working on the online conformation of TV commercials, unlike the editors of music videos, are not working from an edit decision list generated in offline editing. Instead, they start with a footage count list derived from whatever method was employed to transfer the film footage to videotape. They then convert the film footage count to television time, either by hand or by using the automatic conversion features available on some online editing systems.

In conforming commercials, many video editors begin by recording the transferred workprint and the final mixed sound track onto the record master tape (the tape that will be used for recording the final edited commercial during online conformation). They then replace the workprint video with the edited video shot by shot, following the same technique described earlier for music videos.

As discussed earlier in this section, the special effects and video transitions used in TV commercials are added during the online conformation stage. It is a good idea for commercial houses to create storyboards for these effects sequences and to discuss them with the video facility in advance of the online session.

Virtually all commercial houses also rely on the online facility to add the graphics and titles that are such an integral part of many TV spots. To accommodate titles work, video facilities that specialize in commercial projects will usually have several monochrome cameras mounted on adjustable stands. Most also have at least one color camera, so that they can accommodate blue screen cells, company logos, color photographs or the reproduction of any other color graphics.

Electronic graphics, once frowned upon for their stark, electronic look, are becoming much more popular and prevalent in TV commercials. Today, the electronic graphics systems used in many online facilities feature a variety of type fonts and color tones. Newer systems also offer various animation and wipe effects, plus the ability to capture and manipulate individual frames of video information.

Many commercials scheduled for national distribution must eventually be edited to accommodate local dealer tags, local phone numbers, regional price differences and other information that tailors the spots to fit a particular state or region. When this is the case, you will usually be better off creating a generic master that includes slots for the local and regional information. Later, you can use this master to create the specially tailored versions.

CONCLUSION

I began this chapter by stating that the specific steps involved in electronic post-production can differ significantly from project to project, depending on the nature and complexity of the production. For example, in a television series such as "Fame," the film footage for each episode is transferred to video, offline and online editing is performed electronically and the final product is delivered to TV stations on videotape. In contrast, on a feature film such as *Oh God, You Devil,* the workprint dailies are transferred to videotape and a workprint is prepared electronically, but the negative is edited in the conventional film manner and the final product is delivered to theaters as 35mm prints. Finally, in many commercials and music videos, the producers adhere to conventional film procedures through the first stages of production and post-production and then turn to video for final editing and delivery.

Although the stage at which these different types of projects switch to electronic post-production differs, the reason for making the switch is always the same—to save time and money. With production costs continuing to climb and project deadlines getting tighter and tighter, that seems to be reason enough for many film producers. On most projects, then, the question is not really whether a production should go electronic but to what degree and at which stages of the post-production process producers should make the switch.

In the case studies presented in this chapter, I've suggested stages where electronic post-production appears to make the most sense on different types of projects. However, with conditions and technology changing rapidly, producers and editors must analyze each situation for themselves. A good place to start is by reading Chapter 3, where I take a deeper look into some of the technical parameters that have shaped, and that will continue to shape, the development of electronic post-production.

3 Time Code, Technology and the Television Picture

The recipe for success in electronic post-production requires two key ingredients: knowledge and know-how. First, you need to know about the steps and equipment involved in electronic post-production and the various technical factors that determine what you can and can't do in electronic posting. Then, you need to know how to put that technical knowledge to work managing actual editing projects.

In this chapter, I raise and answer many of the fundamental technical questions that come up during electronic post-production, questions such as

- How do you transfer film, shot at 24 frames per second, to a video recording at 30 frames per second?
- How do you convert film footage counts to television time?
- What is SMPTE time code, and how do you use it in electronic post-production?
- What is film time code?

To understand these and other technical matters related to electronic post-production, you first need to understand how a video picture is created on the TV screen.

THE VIDEO IMAGE: TELEVISION FIELDS AND INTERLACED SCANNING

In the United States, the standard television image is created by an electronic beam scanning a picture tube at a rate of 525 lines of picture information 30 times each second. As shown in Figure 3.1, the scanning beam starts at the upper left corner of the picture tube raster and moves from left to right, dropping down a line and then retracing when it reaches the righthand edge of the picture, or raster, area.

But there is an odd twist to this system. Early in the development of television, engineers discovered that, although 30 scans per second produced a picture on the television monitor, the picture contained considerable flicker. To resolve this problem, the designers first tried increasing the scan rate to 60 scans per second. Unfortunately, although this remedy did reduce the flicker, it introduced other problems when the signal was transmitted.

To minimize both sets of problems, television's engineering forefathers developed a system called "interlaced scanning." Interlaced scanning works on the principle of dividing the television frame into two separate fields, each containing 262.5 (half of the original 525) lines of picture information. Field one contains the odd-numbered lines (1, 3, 5, 7, etc.), and field two contains the even numbered lines (2, 4, 6, 8, etc.). Field one is scanned first, producing an image containing only the odd-numbered lines. Then, under the control of vertical synchronization pulses, the electron beam returns to the top of the raster, where it begins scanning the second field, which is made up of the even-numbered lines. By the time the beam reaches the bottom right corner of the raster, it has filled in the rest of the picture (see Figure 3.2). In other words, although the standard U.S. television signal creates a TV picture at the rate of 30 frames per second, each frame contains two interlaced fields, resulting in an actual scanning rate of 60 fields per second.

TRANSFERRING 24-FRAME FILM TO 30-FRAME TELEVISION

As discussed in the previous section, the standard U.S. television signal operates at 30 frames per second. Each television frame is, in turn, composed of two television fields, making for a TV signal scan rate of 60 fields per second. Since film operates at a rate of 24 frames per second, we must find 6 extra frames somewhere to make up the difference during film-to-tape transfer.

This problem was initially solved several years ago, by installing a mechanical device known as an intermittent pulldown (or "3:2 pulldown") assembly in the film projector. The pulldown assembly works by periodically holding a film frame in the projector gate for three television fields' duration (1.5 TV frames) instead of the normal 2 fields' duration (1 TV frame). Keep in mind that 6 television frames equal 12 television fields. It follows, then, that by holding every other film frame in the gate for an extra field duration, the pulldown device produces the 12 extra fields, or the 6 extra frames, needed to make up the difference between film and TV frame rates.

Here is how it works. The pulldown device holds the first frame in the projector gate for 2 TV fields (1 TV frame). Then, the second film frame is held in the projector for 3 television fields. As a result, television frame two consists of 2 fields of the second film frame, and television frame three consists of 1 field of film frame two and 1 field of film frame three. Table 3.1 illustrates the pulldown ratio of

Figure 3.1: Original Television Signal Scanning Method

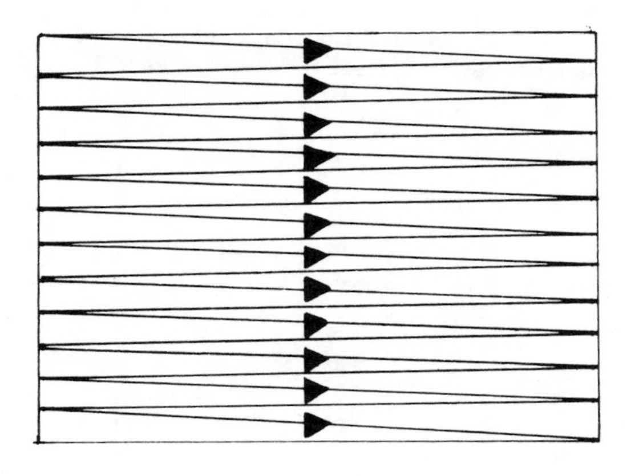

Figure 3.2: Interlaced Scanning

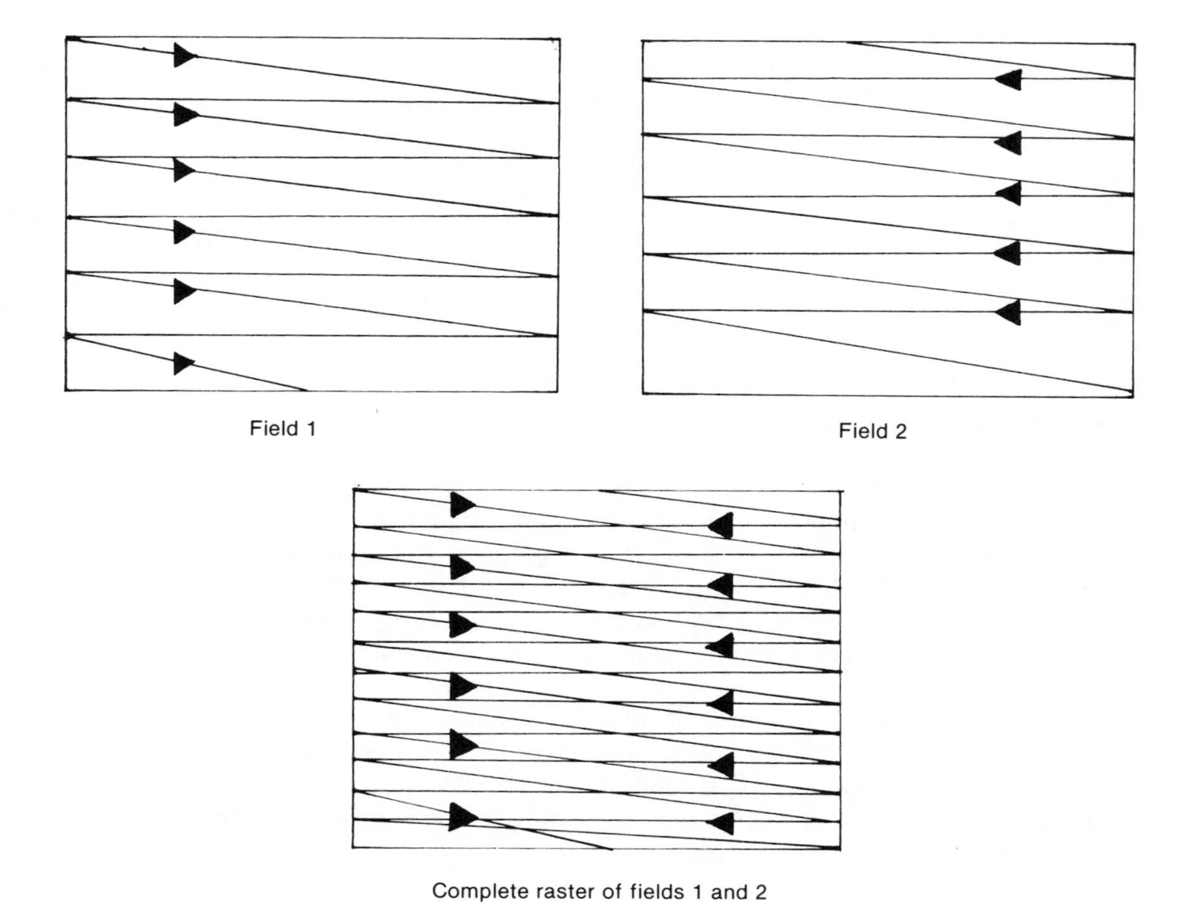

Table 3.1: Pulldown Rates in Film-to-Tape Transfer

Television Frame	Film Frames In Field 1	Film Frames In Field 2
1	1	1
2	2	2
3	2	3
4	3	4
5	4	4
6	5	5
7	6	6
8	6	7
9	7	8
10	8	8
11	9	9
12	10	10
13	10	11
14	11	12
15	12	12
16	13	13
17	14	14
18	14	15
19	15	16
20	16	16
21	17	17
22	18	18
23	18	19
24	19	20
25	20	20
26	21	21
27	22	22
28	22	23
29	23	24
30	24	24

film frames to TV frames during the film-to-tape conversion process—and how it all adds up to 24 film frames filling 30 TV frames each second.

The pulldown process can present problems during editing. For example, if a pulldown occurs in conjunction with a camera cut, the particular television frame involved would contain one field of the outgoing scene and one field of the incoming scene. If this film-to-tape transfer is then edited on videotape, an edit performed on that frame will appear as an annoying "flash frame." Fortunately, this can be easily corrected by adjusting the edit point one frame.

Another problem can appear during "pan scans"—a telecine technique in which only a designated portion of a wide-screen film is transferred to videotape for broadcast on the small-screen medium of television. If the pan starts on a pulldown frame, the transfer will appear as a double image or miscut. This can be remedied by advancing or retarding the start of the pan by one frame.

PRODUCING FILMS AT 30 FRAMES PER SECOND

To avoid some of the difficulties created by the different frame rates of film and video, many producers of television commercials choose to shoot their productions using special cameras that operate at the rate of 30 frames per second. This results in a one-to-one film–video frame ratio, and it eliminates the need for using the intermittent pulldown device during film-to-tape transfer. The higher frame rate also reduces the "grain" that is apparent in many film images, resulting in a much sharper image resolution.

Shooting at 30 frames per second throughout a production, or in selected sequences, allows producers of TV commercials to add extra sparkle to the sponsor's product. Unfortunately, it can also add about 25% extra to the cost of film stock compared to a normal production.

VARIABLE SPEED FILM-TO-TAPE TRANSFER

Several years ago, I watched an American Film Institute tribute to Lillian Gish that included several clips from her early films. In contrast to the usual procedure of transferring film footage at 24 frames per second, the clips from Miss Gish's early films were transferred at 18 frames per second—the same speed at which they were originally shot. The result was stunning. Instead of the jerky motion that we have come to associate with silent films seen on broadcast television, the scenes were fluid and smooth.

The Lillian Gish clips could be seen in their original splendor because the transfer team took advantage of the variable speed feature available on many modern telecine systems. Of course, the variable speed feature is not used solely for transferring classic film footage to videotape. Film shot for special effects sequences is sometimes transferred at above or below normal speed, as are film sequences that must be compressed to fit into television time slots.

In fact, entire feature films sold for TV broadcasting are sometimes transferred at a faster than normal rate. By speeding up the transfer rate, TV stations and the broadcast networks can often squeeze films to fit into TV time slots without cutting a single scene. For example, when transferred at 28 frames per second, a feature film that normally runs for 115 minutes would end up running for only 98.5 minutes. The result is time savings with very little noticeable difference in the on-screen action. However, to keep the sound normal, a sound pitch correction device must be used during the transfer.

Along with these benefits, variable speed transfer systems can open a number of new production options and opportunities for film producers and corporate media specialists—especially when used in combination with the new film-to-video controllers described in Chapter 4.

CONTRAST RATIOS: TV VERSUS FILM

When transferring film to videotape, it is important to remember that film and television have very different contrast ranges (the range between the lightest and darkest picture elements). It is not uncommon to hear movie producers complain that film footage looks more crisp and vibrant during a production screening than it does after transfer to videotape. What the producers mean is that the videotape version of the film doesn't seem to feature the same degree of separation between the shades of gray in the image—and they are right.

The fact is, film is shot at contrast ranges of 100:1 or better, whereas the TV contrast ratio falls somewhere between 15:1 and 30:1. This means that, with less separation between the lightest and darkest picture elements, the TV image simply can't reproduce all the subtle shades and shadows present in the film image. As a result, transferring film to videotape usually results in a loss of detail in low-light areas or a washing out of detail in high-light areas, depending on how the telecine (film-to-tape) system is adjusted.

THE TELEVISION ASPECT RATIO

Film professionals involved in electronic post-production must also be aware of the television aspect ratio, the standard ratio of picture height to width. In the American NTSC (National Television Standards Committee) 525/60 TV standard, the aspect ratio is 3:4 (see Figure 3.3). That is, for every 3 inches of height, the scanned area measures 4 inches across (9″ × 12″, 12″ × 16″, etc.). However, due to transmission losses and TV receiver misadjustments, most home TV sets do not display the entire scanned picture.

To avoid recording information in the cutoff zone around the edges of the TV picture, video engineers have developed two related standards: the safe title area and the safe action area (also shown in Figure 3.3). The safe title area corresponds to the film "8 field" (80% of the picture area), and it marks the boundary of the zone safely within the readable scanning area of home television receivers. The safe action area corresponds to the film "9 field" (90% of the picture information)—the absolute boundary for ensuring that the action displayed on a TV camera or monitor is not cut off by the home TV screen. Both boundary areas are of critical concern when an editor is working with still or motion graphics.

CONVERTING FILM FOOTAGE TO TELEVISION TIME

Editors often need to know how long a given length of film will run once it is transferred to videotape. Because film length is usually measured and listed in

Figure 3.3: Safe Title and Safe Action Boundaries

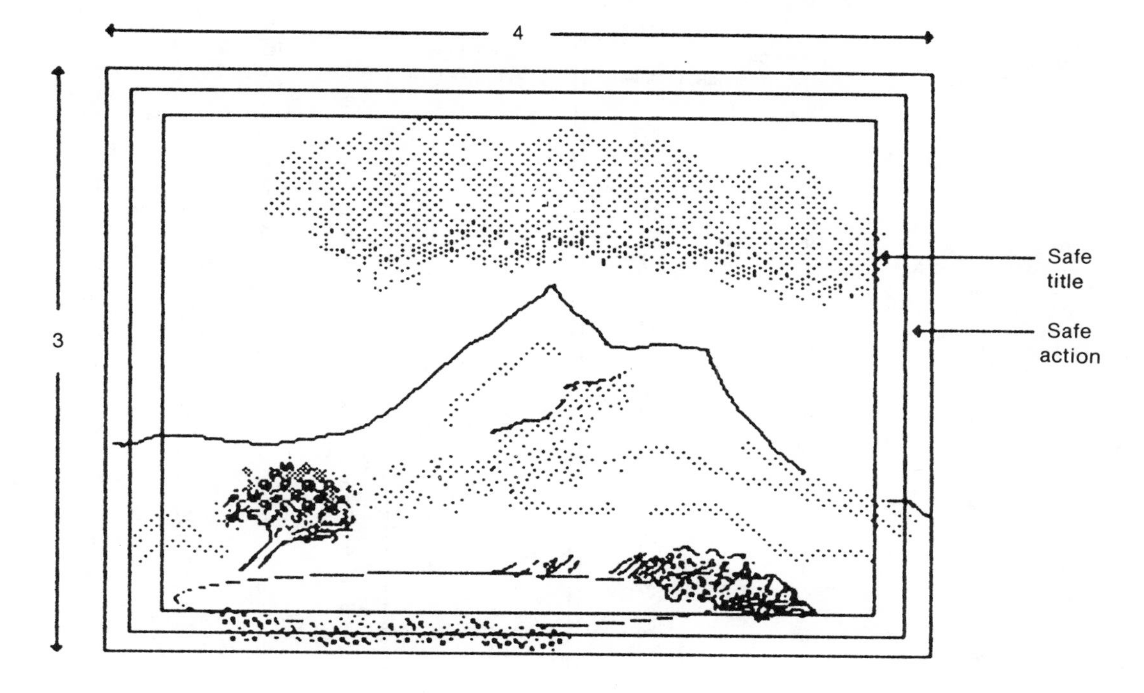

"footage," I've supplied two tables for converting these footage measurements to videotape running time. Table 3.2 is for use with 16mm film; Table 3.3 for use with 35mm film.

To use the tables, find the length of film footage in the left columns, then note the television time equivalent from the corresponding righthand column. For example, if you had a 16mm film clip that totaled 23 feet 26 frames, you would look up the foot and frame numbers in the left hand column of Table 3.2. Then, looking to the righthand column, you would note the following:

23 feet of 16mm film = 38 seconds and 9 frames of TV time
26 frames of 16mm film = 1 second and 2 frames of TV time

By adding the two conversions, you get a total of 39 seconds and 11 frames of television time from 23 feet and 26 frames of 16mm film.

SMPTE/EBU TIME CODE

To understand electronic post-production, filmmakers must have a basic understanding of SMPTE/EBU (European Broadcasting Union) time code (commonly called "SMPTE code"). As Figure 3.4 shows, SMPTE time code is indexed in hours, minutes, seconds and frames (30 frames per second in the U.S. NTSC standard, and 25 frames per second in the European PAL standard). NTSC time

Table 3.2: Conversion of Film Footage to TV Time for 16mm Film

16mm Film		Television				16mm Film		Television			
Feet	Frames	Hours	Minutes	Seconds	Frames	Feet	Frames	Hours	Minutes	Seconds	Frames
0	0	0	0	0	0	0	38	0	0	1	17
0	1	0	0	0	1	0	39	0	0	1	19
0	2	0	0	0	2	1	0	0	0	1	20
0	3	0	0	0	4	2	0	0	0	3	10
0	4	0	0	0	5	3	0	0	0	4	30
0	5	0	0	0	6	4	0	0	0	6	20
0	6	0	0	0	7	5	0	0	0	8	10
0	7	0	0	0	9	6	0	0	0	9	30
0	8	0	0	0	10	7	0	0	0	11	20
0	9	0	0	0	11	8	0	0	0	13	10
0	10	0	0	0	12	9	0	0	0	14	30
0	11	0	0	0	14	10	0	0	0	16	20
0	12	0	0	0	15	11	0	0	0	18	9
0	13	0	0	0	16	12	0	0	0	19	29
0	14	0	0	0	17	13	0	0	0	21	19
0	15	0	0	0	19	14	0	0	0	23	9
0	16	0	0	0	20	15	0	0	0	24	29
0	17	0	0	0	21	16	0	0	0	26	19
0	18	0	0	0	22	17	0	0	0	28	9
0	19	0	0	0	24	18	0	0	0	29	29
0	20	0	0	0	25	19	0	0	0	31	19
0	21	0	0	0	26	20	0	0	0	33	9
0	22	0	0	0	27	21	0	0	0	34	29
0	23	0	0	0	29	22	0	0	0	36	19
0	24	0	0	0	30	23	0	0	0	38	9
0	25	0	0	1	1	24	0	0	0	39	29
0	26	0	0	1	2	25	0	0	0	41	19
0	27	0	0	1	4	26	0	0	0	43	9
0	28	0	0	1	5	27	0	0	0	44	29
0	29	0	0	1	6	28	0	0	0	46	19
0	30	0	0	1	7	29	0	0	0	48	9
0	31	0	0	1	9	30	0	0	0	49	29
0	32	0	0	1	10	31	0	0	0	51	18
0	33	0	0	1	11	32	0	0	0	53	8
0	34	0	0	1	12	33	0	0	0	54	28
0	35	0	0	1	14	34	0	0	0	56	18
0	36	0	0	1	15	35	0	0	0	58	8
0	37	0	0	1	16	36	0	0	1	00	00

code readouts range from 00:00:00:00 to 23:59:59:29, recycling at each 24-hour interval.

Methods of Recording Video Time Code

There are two ways to record SMPTE time code on videotape. In the most common method, known as longitudinal or "serial" time code (LTC), the time code is recorded on the videotape along one audio track. In the second method, known as vertical interval code (VITC), the code is recorded in the vertical interval (an unused portion of the TV signal) during film-to-tape transfer. In both cases, SMPTE code is recorded as a binary (digital) signal designed to generate a complete line of information twice during each television frame (or one complete line of information per field).

Table 3.3: Conversion of Film Footage to TV Time for 35mm Film

35mm Film		Television				35mm Film		Television			
Feet	Frames	Hours	Minutes	Seconds	Frames	Feet	Frames	Hours	Minutes	Seconds	Frames
0	1	0	0	0	1	39	0	0	0	26	0
0	2	0	0	0	2	40	0	0	0	26	20
0	3	0	0	0	4	41	0	0	0	27	10
0	4	0	0	0	5	42	0	0	0	28	0
0	5	0	0	0	6	43	0	0	0	28	20
0	6	0	0	0	7	44	0	0	0	29	10
0	7	0	0	0	9	45	0	0	0	30	0
0	8	0	0	0	10	46	0	0	0	30	20
0	9	0	0	0	11	47	0	0	0	31	10
0	10	0	0	0	12	48	0	0	0	32	0
0	11	0	0	0	14	49	0	0	0	32	20
0	12	0	0	0	15	50	0	0	0	33	10
0	13	0	0	0	16	51	0	0	0	34	0
0	14	0	0	0	17	52	0	0	0	34	20
0	15	0	0	0	19	53	0	0	0	35	10
1	0	0	0	0	20	54	0	0	0	36	0
2	0	0	0	1	10	55	0	0	0	36	20
3	0	0	0	2	0	56	0	0	0	37	10
4	0	0	0	2	20	57	0	0	0	38	0
5	0	0	0	3	10	58	0	0	0	38	20
6	0	0	0	4	0	59	0	0	0	39	10
7	0	0	0	4	20	60	0	0	0	40	0
8	0	0	0	5	10	61	0	0	0	40	20
9	0	0	0	6	0	62	0	0	0	41	10
10	0	0	0	6	20	63	0	0	0	42	0
11	0	0	0	7	10	64	0	0	0	42	20
12	0	0	0	8	0	65	0	0	0	43	10
13	0	0	0	8	20	66	0	0	0	44	0
14	0	0	0	9	10	67	0	0	0	44	20
15	0	0	0	10	0	68	0	0	0	45	10
16	0	0	0	10	20	69	0	0	0	46	0
17	0	0	0	11	10	70	0	0	0	46	20
18	0	0	0	12	0	71	0	0	0	47	10
19	0	0	0	12	20	72	0	0	0	48	0
20	0	0	0	13	10	73	0	0	0	48	20
21	0	0	0	14	0	74	0	0	0	49	10
22	0	0	0	14	20	75	0	0	0	50	0
23	0	0	0	15	10	76	0	0	0	50	20
24	0	0	0	16	0	77	0	0	0	51	10
25	0	0	0	16	20	78	0	0	0	52	0
26	0	0	0	17	10	79	0	0	0	52	20
27	0	0	0	18	0	80	0	0	0	53	10
28	0	0	0	18	20	81	0	0	0	54	0
29	0	0	0	19	10	82	0	0	0	54	20
30	0	0	0	20	0	83	0	0	0	55	10
31	0	0	0	20	20	84	0	0	0	56	0
32	0	0	0	21	10	85	0	0	0	56	20
33	0	0	0	22	0	86	0	0	0	57	10
34	0	0	0	22	20	87	0	0	0	58	0
35	0	0	0	23	10	88	0	0	0	58	20
36	0	0	0	24	0	89	0	0	0	59	10
37	0	0	0	24	20	90	0	0	1	0	0
38	0	0	0	25	10						

SMPTE code provides eight optional slots ("user bits") that can be used for inserting any information deemed necessary during production. These slots can accommodate enough digital data for four alphabetical characters, eight numerical characters or a combination of the two. (See Figure 3.4.)

Figure 3.4: Typical Window Dupe Recording with the Addition of User Bit Time Code and Longitudinal Time Code

Photo by Darrell R. Anderson

Producers usually use the slots to insert reference time code from the master playback track during musical productions, but they may also include information about one or more of the following: reel numbers, take numbers, scene numbers and production dates. The information is entered by front panel controls on the time code generator used during production or the film-to-tape transfer.

Operating Modes for SMPTE Time Code

There are two operating modes for 30-frame-per-second SMPTE time code: non-drop-frame and drop-frame. Non-drop-frame time code (also called "color time code") is a continuously counting time code that indicates hours, minutes, seconds and frames and that resets at each 24-hour interval. Drop-frame time code (also called "clock time code") works in much the same way, but with one significant difference. It drops, or skips, frames at designated time intervals so that the time code will maintain real-time accuracy.

As discussed earlier in the section on interlaced scanning, the NTSC standard television signal operates at 59.94 Hz, not the 60 Hz at which clocks operate. As a result, the clock and the TV signals are actually running at slightly different speeds. This leads to the sort of problem you would encounter if you started two ¼-inch audio machines at the same time but ran them at slightly different speeds. After a short time, they would begin running out of sync, and at a gradually expanding rate.

A similar effect occurs with SMPTE time code and the TV signal. The two may start in sync with the clock, but, because of the speed difference just described, there will be an expanding elapsed time difference between the clock and the time code readings. As Figure 3.5 illustrates, the actual difference amounts to 108 frames per hour, or 86 seconds for each 24-hour period. To correct for this time difference, drop-frame time code is designed to skip two frames a minute, with the exception of every tenth minute.

For a number of reasons, time code editing with noncomputerized editing systems is much simpler using non-drop-frame code. While the non-drop-frame code is not real-time accurate, it is considerably easier to calculate manually than

Figure 3.5: Drop-Frame versus Non-Drop-Frame Time Code Error Factor

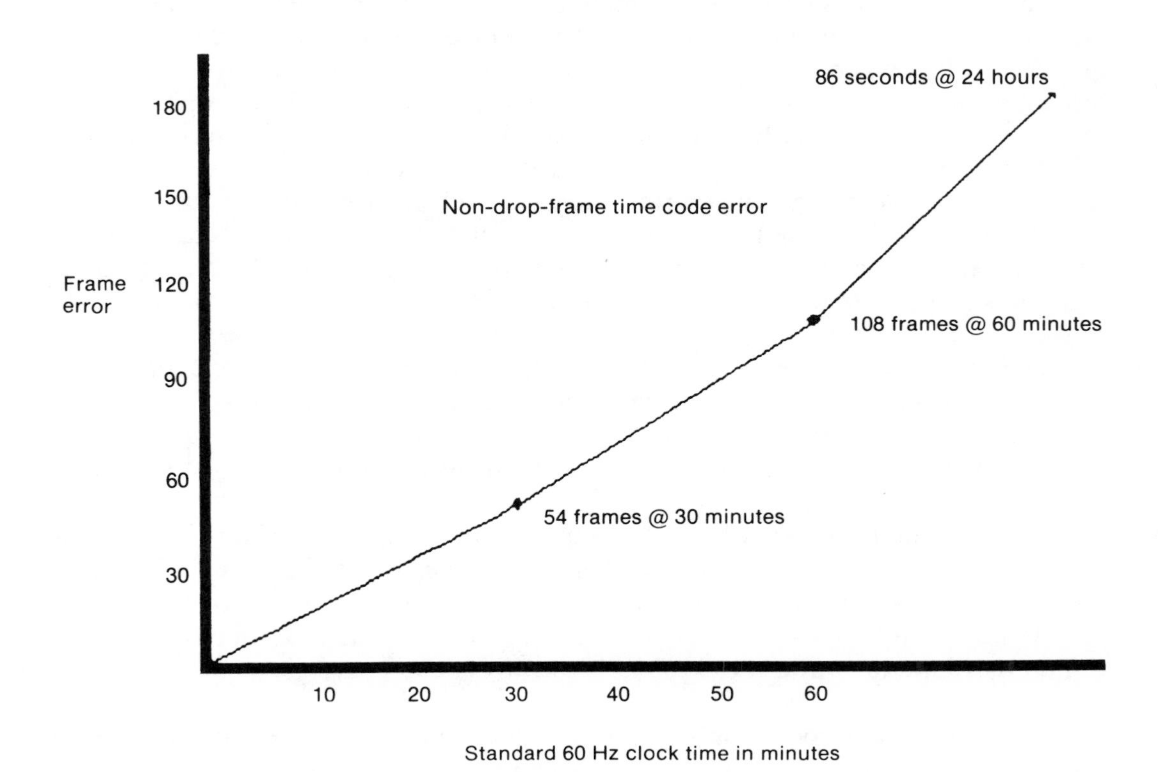

drop-frame code. Just keep in mind that, at the end of a 30-minute project, the elapsed time on the program will be one second plus 24 frames longer than the reading on the time code display.

An industry rule of thumb is that, for each half-hour of programming, a production using non-drop-frame code is actually two seconds longer than the time code indicates. Therefore, if the program is designed to be 28 minutes and 39 seconds long, the calculated non-drop-frame time should equal only 28 minutes and 37 seconds.

FILM TIME CODE

As the case studies in Chapter 2 suggest, filmmakers and video producers have become increasingly interested in combining film and video production and post-production techniques. For example, while many filmmakers have begun to transfer and edit their projects on videotape, many video producers have begun adopting film-style lighting and shooting techniques. As a result, film and video personnel who once worked in totally separate worlds are now working with, and learning from, each other.

As part of the trend toward combining film and video production, many major motion picture studios, including Paramount and Universal, have begun investing in video facilities and electronic production techniques. For example, Paramount has converted several film sound stages to video production facilities, where it now produces shows such as "Entertainment Tonight," "Family Ties," "Webster" and "Brothers." Like Paramount, Universal has also converted several sound stages, and it is experimenting with various forms of electronic editing. In addition, Universal is playing a significant role in one of the most critical stages in the development of electronic post-production: the struggle to produce a practical method for using time code in film production.

Unfortunately, Universal and other groups involved in this struggle have been hampered by the fact that production procedures vary widely from one film studio to the next. This makes it difficult for any single organization to establish and enforce the sort of specifications necessary to develop a film time code standard.

Efforts to develop a time code standard for film date from the late 1960s, when the manufacturers of video equipment first established standards for a TV time code. The goal of TV and film time code is the same: to allow production and post-production professionals to search for and synchronize footage automatically, with frame-by-frame accuracy.

When TV time code was being developed, filmmakers were using a time-tested method to record and locate film segments, the "edge number system." In the edge number system, frame numbers are engraved along the edge of the film.

Unfortunately, since edge numbers must be logged and cataloged by hand, the edge number system does not permit the kind of automatic cueing and editing that is the primary benefit of TV time code.

Aware of this limitation, European engineers went to work on developing a method of time-coding film. Their early efforts culminated in 1976, when the EBU Four-Bar Time Code was introduced, one of the earliest incarnations of film time code.

EBU Optical Four-Bar Film Time Code

In the EBU Four-Bar Time Code system, digital codes resembling today's supermarket bar codes were burned into the edge of the film. Unfortunately, the EBU code contained only four "bits" of information per film frame (compared to the 80 bits per frame contained in SMPTE video time code). Because of this limitation, EBU time code lacked the capacity and flexibility necessary to make it an effective production and post-production tool. As a result, it never really caught on in the film industry, despite the determined efforts of its designers to improve it through many modifications.

SMPTE .8mm Optical Bar Film Time Code

SMPTE .8mm Optical Bar Film Time Code, another application of bar code technology, was originally developed in the late 1970s to facilitate the captioning of filmed programming intended for TV distribution. The process involved recording a continuous 80-bit SMPTE time code signal on an .8mm-wide strip alongside the sound track area of 16mm positive film.

Although the SMPTE .8mm system worked well under laboratory conditions, problems popped up when film teams tried to use it in actual production situations. The process was plagued by exposure errors, unreliable contrast values, scratches induced by the film rollers, and other problems that rendered the delicate code unreadable. As a result, the advocates of film time code quickly began searching for another, more reliable process.

SMPTE LTC 2mm Optical Bar Film Time Code

Another version of SMPTE Bar Code, 2mm Optical Bar Film Time Code, was developed in 1982. In this system, a 2mm-wide stream of 80-bit time code is recorded in the sound track area of 16mm film.

Unlike the .8mm bar code just described, the 2mm system was dust and scratch resistant. However, because of the positioning of the bar code, 2mm time code could only be used with monoperf film (film with sprocket holes along only one side of the film). Because many film projects use doubleperf film, designers proposed modifying the code so the bit stream would fit between the perforation holes on the film. Unfortunately, although this change would have allowed filmmakers to use 2mm time code with doubleperf film, it created problems similar to those that plagued the .8mm process, including readability problems that resulted from reduced legibility.

Aaton Film Data Track (FDT) Optical Film Time Code

In the mid-1980s, Aaton Camera, Inc. introduced still another method of recording time code on film. Aaton's Film Data Track (FDT) system uses an array of light-emitting diodes (LEDs) mounted inside a 16mm or 35mm camera to expose SMPTE time code on the film negative.

The time code track resembles a 7-by-13-square checkerboard. Because the exposed checkerboard surface is 10 times larger than the surface in linear bar code, the FDT approach is far less susceptible to the contrast variations, scratches and reader-head alignment problems that have afflicted other optical time code systems.

The heart of the system is a master clock generator that Aaton calls the Origin C. Resembling a pocket calculator, the Origin C is used to initialize the camera and sound recorder generators at the start of each day's shooting. By pressing numbers on the Origin C, crew members can enter the time, date and production code.

On 16mm productions, the FDT time code is exposed between the perforations on the claw side of the film. On 35mm productions, the code is recorded outside the perforations. In both cases, there is no contact between the LED head and the film surface, reducing the chance that the negative will be scratched during the coding process. Aaton is also planning to add a small computer keyboard to the system so that film producers can enter production notes as part of the time code during filming.

The FDT system has two operating modes: the "code" mode and the "clear" mode. If the production is to be edited electronically, the film crew uses the code mode to record SMPTE time code on 19 of the 24 frames exposed each second, with the other five frames used for exposing numbers that can be read by the naked eye. If the project is to be edited using conventional film techniques, the crew uses the clear mode to insert an "address" that can be read by the naked eye for every second of exposed film.

Datakode Magnetic Film Time Code

In 1982, Eastman Kodak Co. introduced a line of products based on a magnetic method of recording time code on the film negative. The system, called

Datakode Magnetic Film Time Code, places a thin layer of magnetic oxide across the entire base of the nonemulsion side of the film (see Figure 3.6). During shooting, a time code generator and mechanical recording head located in or on the camera combine to record the time code on the magnetic surface. This camera time code is synchronized with the code being recorded on track two of the production audio recorder, to facilitate synchronization and sound post-production.

Although it is functionally transparent to the camera, the magnetic coating does impart some additional density to the film. As a result, filmmakers who plan on using the Datakode system should first check the tolerances of their equipment and then check with Kodak to receive the exact Datakode density specifications.

Although the Datakode process holds much promise, Kodak engineers are still working on some lingering problems with the system. For example, at this writing, the Datakode process requires the record head to maintain physical contact with the film, introducing the possibility of scratching the negative. As a partial solution to this problem, Kodak chose to relegate the time code information to two .8mm strips recorded on either edge of the film. This reduces the chance that valuable information will be harmed if the recording head scratches the surface of the negative.

Kodak engineers are also working to correct early difficulties with maintaining accurate time code readings across film splices. In addition, Kodak has been busy working on fixes for various hardware-related problems. Those problems include the need to install separate record and confidence heads in each film magazine used in instant-change cameras, and the need to make adjustments to

Figure 3.6: Cross-Section of Motion Picture Film Showing the Emulsion Layers and Datakode Magnetic Control Surface

Yellow-dye picture layer
Magenta-dye picture layer
Cyan-dye picture layer
Film support
Magnetic control surface
Antihalation layer

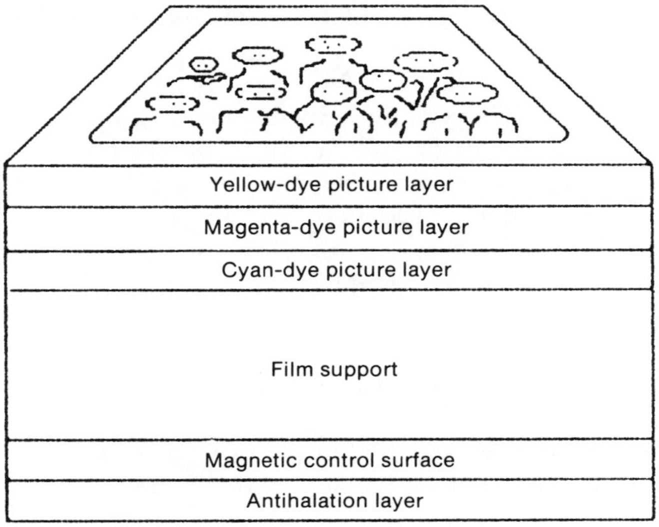

the time code to compensate for the mechanical differences between various pieces of production equipment.

Why Bother with Film Time Code?

Given all the obstacles in the way of developing a truly effective film time code system, why bother with developing a film time code at all? The answer, of course, is that the many equipment companies and film studios involved in the struggle are betting that the film time code will eventually pay off big.

Once a film time code standard is set, filmmakers will quickly realize significant time and cost savings. For example, with a standardized time code system, producers will no longer have to mark the beginning of each scene with the familiar film slate used to identify the shoot and facilitate sound synchronization. Along the same lines, filmmakers working on documentary-style or multicamera concert productions will no longer have to risk losing a critical shot because they must pause to shoot a clapboard or synchronized clock display. Production assistants will also benefit from a film time code standard, if only because items such as scene and take numbers, shooting dates, roll numbers and other key production data can be recorded quickly and securely using the user bit feature of SMPTE time code. Cataloging production information in this way makes it easy to retrieve specific shots and scenes using only a brief description of the segment or a time code number.

In post-production, film time code will allow crew members to automate the process of synchronizing dailies and to eliminate the need to strike a 35mm magnetic film sound track when synchronizing the ¼-inch sound track during the film-to-tape transfer process. In addition, on film projects that are shot with a "video assist" (a video camera and VCR operator recording the action along with the film crew), the film time code will also appear in the videotape picture, allowing the producer and director the luxury of previewing scenes and selecting editing points on location.

Finally, film time code also has a number of implications for film distribution and display. For example, film distribution facilities will be able to use time code to check for breaks, splices and missing segments on distribution prints. Time code can also be used to automate the process of changing projectors, opening curtains and dimming lights in movie theaters, as well as to help discourage pirating by providing each release print with an identification code.

CONCLUSION

When a film project runs into technical problems during electronic post-production, producers will inevitably blame the problems on that mysterious demon "electronics." In this chapter, I have discussed several of the key technical issues raised by electronic post-production, in the hope that producers and film crew members will be able to recognize and dispel the demons before they do their damage.

Much of the chapter was devoted to discussing issues related to television time and video and film time code. I began by describing how one frame of TV image is formed by an electron beam scanning two fields of video information, and how this interlaced scanning system produces 30 frames of video per second. I also discussed how intermittent pulldown assemblies first solved the problem of "finding" the six extra frames per second necessary to match film time to video time during film-to-tape transfer.

During film-to-tape transfer, film professionals should also remember to account for the differences between the contrast ranges and aspect ratios of film and video. As I pointed out, compared to film, video features a much narrower contrast range and a more restrictive aspect ratio.

I ended the chapter with brief analyses of TV and film time code. Two types of TV time code are now standardized under SMPTE specifications: drop-frame (or "clock accurate") time code and non-drop-frame time code. These two types of time code are now used throughout the television industry as a means of automating a variety of production and post-production processes, including the online editing process.

Film time code has yet to gain such widespread acceptance, primarily because no single system has emerged as the industry standard. However, recent technical advances promise to move film time code out of its experimental stage, and standardization will surely follow. A fully tested, standardized film time code system promises to open up many new production opportunities and post-production options for filmmakers—particularly when it is used in conjunction with the new telecine systems described in the next chapter.

4 Understanding the Telecine Process

Two categories of film material are used in television. First, there are theatrical, documentary and instructional films that were originally designed to be projected on a movie screen but have now been slated for broadcast, cable or videotape distribution. Ideally, the broadcast or video image should match the projected image as closely as possible. Producing this close match is the relatively straightforward job of the *broadcast* telecine system.

Second, there are films and film footage produced specifically for television. As described in Chapter 2, film material shot for a TV production is usually intercut (edited) with film from the same production session, film from another production session or a film library, or images generated by a video camera. Proper reproduction of this film is the job of the *post-production* telecine system.

This chapter examines how a post-production telecine system works. Originally, telecine systems were fairly simple devices designed to do just one task—transfer film for TV transmission. In recent years, however, telecine systems have become increasingly elaborate, and the role that they play in electronic post-production has become increasingly complex.

WHAT IS A TELECINE SYSTEM?

Today, most major video post-production facilities and virtually all broadcast television studios are equipped with a telecine system of some sort. Stated simply, a telecine is an optical-electronic system used to reproduce film or slides for television. To reproduce this material faithfully, the optical film or slide image must be converted to a high-quality color video signal.

For film material, the basic telecine conversion process works like this. First, the film frames are passed through a film gate in the telecine system. Light is projected through the gate, and a special prism separates the projected film image into its red, green and blue color components. Then, the separated color signals are processed through various electronic sensors, amplifiers and color correction circuits. Finally, before exiting the system at the telecine's video output jack, the separate signals are combined into a composite color video signal. (See Figure 4.1.)

Currently, there are three types of post-production telecine systems: the camera tube–type, the flying spot scanner and the line array (or "CCD") telecine system.

CAMERA TUBE–TYPE TELECINE

Camera tube–type telecine systems have been around since the earliest days of color television, and they are still being used in many broadcast, industrial and educational television facilities (see Figure 4.2). In this type of telecine, 16mm or 35mm footage is projected directly into the lens of a TV camera equipped with three or four light-sensitive storage-type pickup tubes. The three key components of the system are the film projector, the multiplexer unit and the TV camera.

The Projector

The projector used in the camera tube–type telecine is a conventional film projector equipped with an intermittent pulldown assembly. As explained in Chapter 3, the pulldown assembly holds every other film frame in the gate for three television fields, as a way to compensate for the difference in the frame rates of film and television.

When the pulldown assembly is operating, the telecine system forms film loops at the entrance and exit guides of the film gate. This helps isolate the film

Figure 4.1: A Basic Telecine System

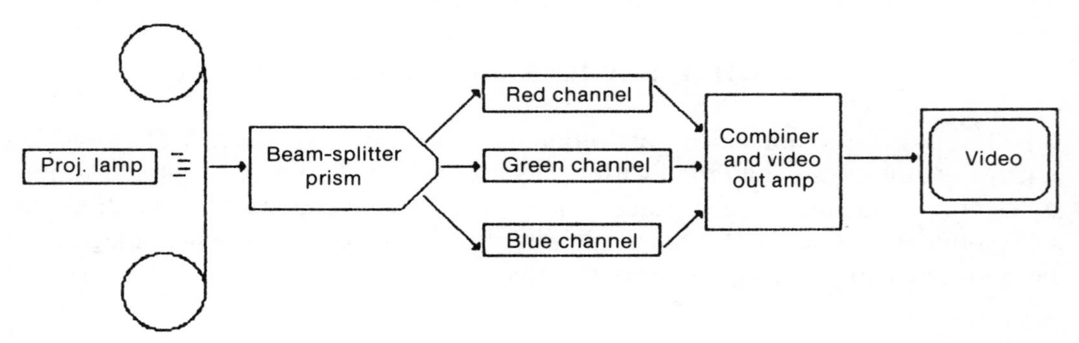

Figure 4.2: RCA TK-29B High-Performance Broadcast Telecine Camera

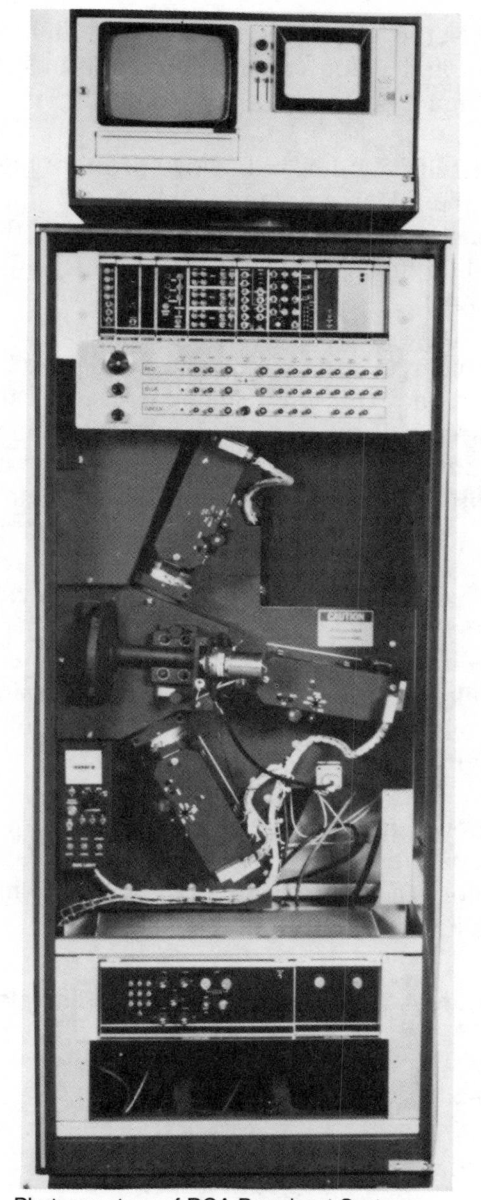

Photo courtesy of RCA Broadcast Systems.

from the staccato motion of the pulldown assembly, reducing film wear and the risk of damage during the projection process.

Most camera tube–type telecine systems are equipped with both 16mm and 35mm projectors, at least one slide projector and a multiplexer unit. As explained below, the multiplexer unit allows the telecine operator to switch between the film and slide sources during the system's operation.

Multiplexer Unit

The multiplexer is the telecine's traffic controller. Together with the film projectors, slide drum and video camera, it constitutes the "film island" of the telecine system (see Figure 4.3).

The multiplexer itself is composed of two or more high-quality mirrors and a gear assembly. When the telecine operator presses the appropriate buttons on the control panel, the gear assembly changes the angle of the mirrors. This in turn changes the film or slide image that is being reflected into the lens of the video camera. (See Figure 4.4.)

On modern telecine systems, this changeover occurs in one-fifth of a second (six telecine frames). As a result, the telecine operator can switch among the film projectors and slide drums during the transfer without interrupting the flow of the program. In addition, audio follow relays built into the multiplexer allow the film sound to "follow along" with each change in film projector.

The Camera

The cameras used in camera tube–type telecine systems are usually equipped with vidicon-type pickup tubes (see Figure 4.5). Measuring approximately one inch in diameter and only six inches in length, vidicon tubes combine high quality and high light sensitivity in a relatively small package.

As Figure 4.6 shows, most cameras used in camera tube–type telecine systems feature three separate vidicon tubes—one each for sensing red, green and blue, the primary colors in video. Some cameras also include a fourth tube for generating a separate luminance (brightness) signal.

Figure 4.3: A Basic Broadcast "Film Island"

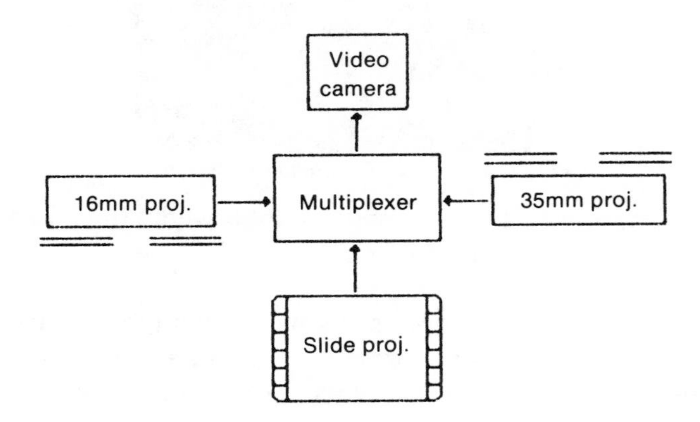

Figure 4.4: RCA TP-55B Telecine Camera Multiplexer

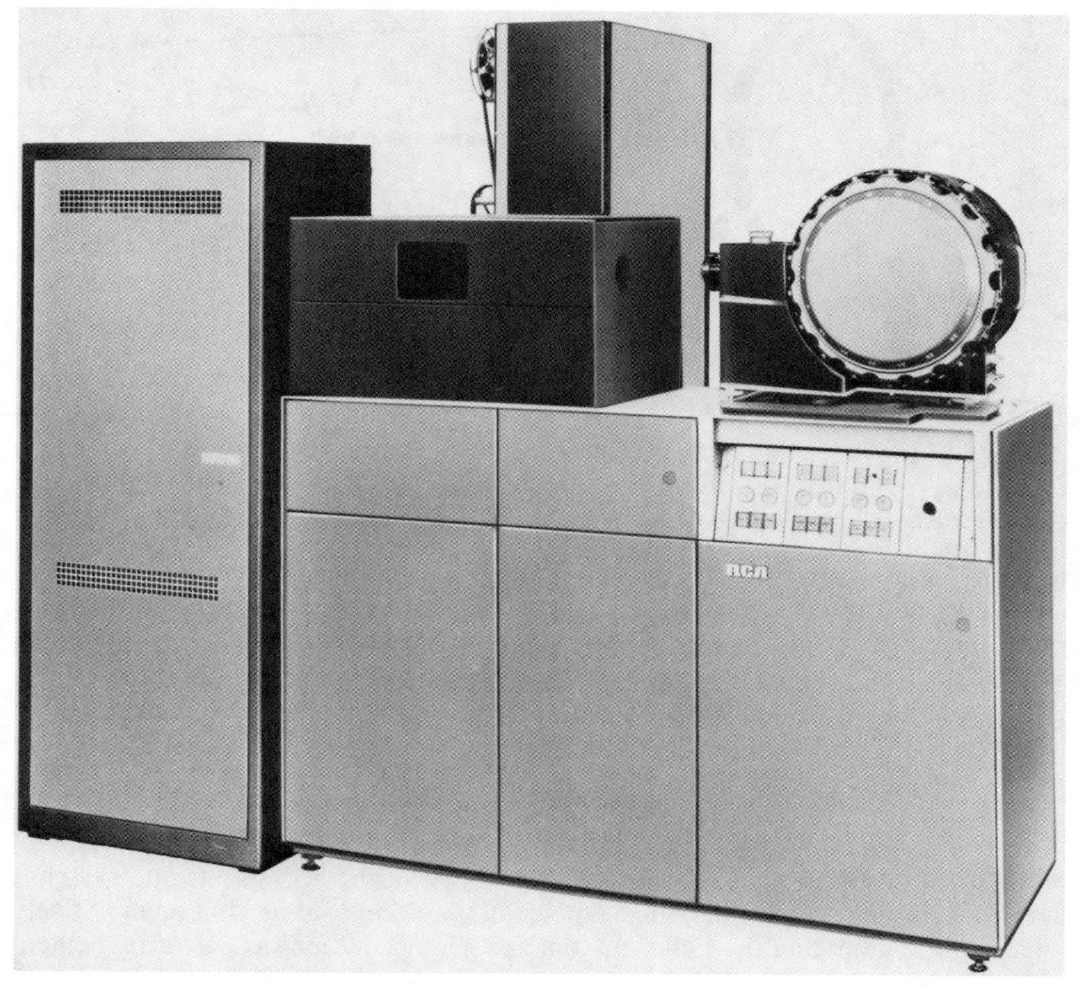

Photo courtesy of RCA Broadcast Systems.

In the telecine process, the projected film image enters the camera lens and strikes beam-splitting (dichroic) mirrors. The dichroic mirrors split the image into its primary colors, reflecting each color through filters and onto the appropriate vidicon tube.

Inside the front face of the vidicon tube is the "signal electrode"—a transparent, conductive coating. The electrical resistance of the coating changes when exposed to the light that enters the vidicon tube. At the base of the tube is an electron gun that "shoots" a narrow, focused beam of electrons at the photoconductive surface of the signal electrode.

As light enters the vidicon tube, an electrical charge builds up on the photoconductive surface of the signal electrode. The brighter the light, the higher the

Figure 4.5: A Basic Vidicon Camera Tube

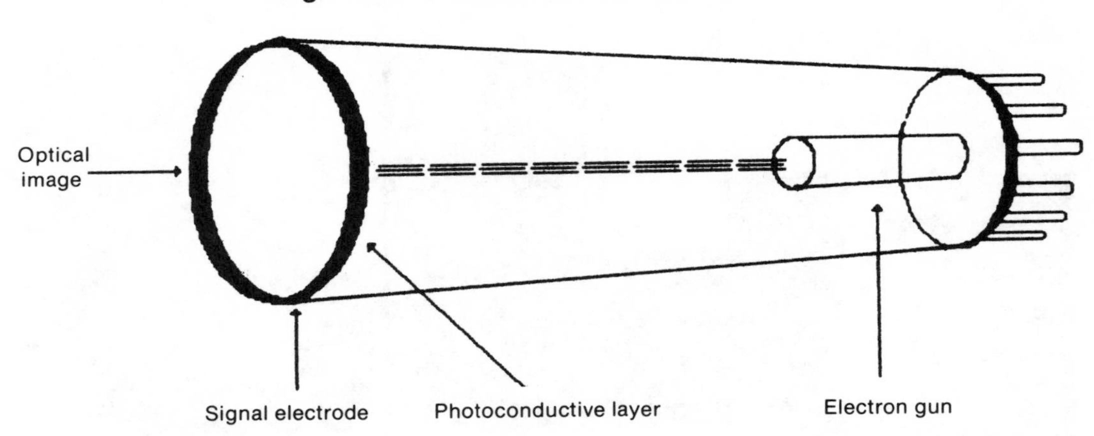

charge. As the electron beam scans the signal electrode, the camera senses these variations, and it uses this information to generate a video signal. The amplitude of the video signal varies in direct relation to the brightness of the optical image that is striking the face of the vidicon tube.

Through this process, the camera generates three separate video signals—one for each primary color. These separate channels are then amplified and processed. Finally, the three channels are combined to produce the video output signal that is transmitted or recorded on videotape.

FLYING SPOT SCANNER TELECINE

Currently, flying spot scanners are the "state of the art" in telecine systems. As shown in Figure 4.7, the flying spot scanner system is housed in one relatively small unit, eliminating the bulky multiple projector and multiplexer arrangement characteristic of camera tube–type telecines. To switch among 35mm film, 16mm film, Super 8mm film and slides, the telecine operator simply changes the size of the filmgate.

In flying spot systems, a single cathode ray tube provides the light used to scan the moving film. An objective lens focuses the CRT raster image on the film as it moves through the gate, and the light beam passing through the film is modulated by the density and color of the image. Next, the modulated optical signal travels to dichroic mirrors, which split the signal into its red, blue and green components. The split signals are then passed onto the red, green and blue photo multiplier tubes, before being amplified, processed and combined into the video output signal. (See Figure 4.8.)

Because the cathode ray tube uses a very fine grain phosphor, the flying spot scanner generates a higher resolution image than that generated by a camera tube–

type system. In the flying spot system, each particle of phosphor in the cathode ray tube acts as a light emitter for a discrete area of the film frame. As a result, the system detects a maximum amount of image information. Also, in flying spot scanners, there is no image lag to smear the picture, as there sometimes is in camera tube–type systems. However, due to the CRT phosphor decay time, flying spot scanner systems do suffer from a slight, but correctable, "afterglow" effect.

Flying spot scanners also benefit from a number of improvements in film handling. For example, because its capstan drive servo system eliminates the need for mechanical film drive sprockets, the flying spot telecine can get up to operational speed—or come down to a full stop—almost instantly. The absence of sprocket claws also allows the flying spot system to run damaged footage, while reducing the risk of film tearing during the transfer process. In addition, the design of the synthetic capstan, coupled with the fact that the capstan is the only part to contact the film area, drastically reduces the chance that the film will be scratched during transfer. Finally, the absence of sprocket drives, mechanical gearboxes and other clanky parts offers flying spot scanners one other advantage over older, more mechanical systems—relatively quiet operation.

Along with these advantages, flying spot scanners are very versatile machines. Some of the features and options offered by the typical flying spot system are described below.

Variable Speed Shutter Control

With its variable speed shutter control, the flying spot scanner can shuttle film at 10 times (for 35mm film) and 20 times (for 16mm film) normal projection speed.

Figure 4.6: A Typical Three-Tube Telecine Camera

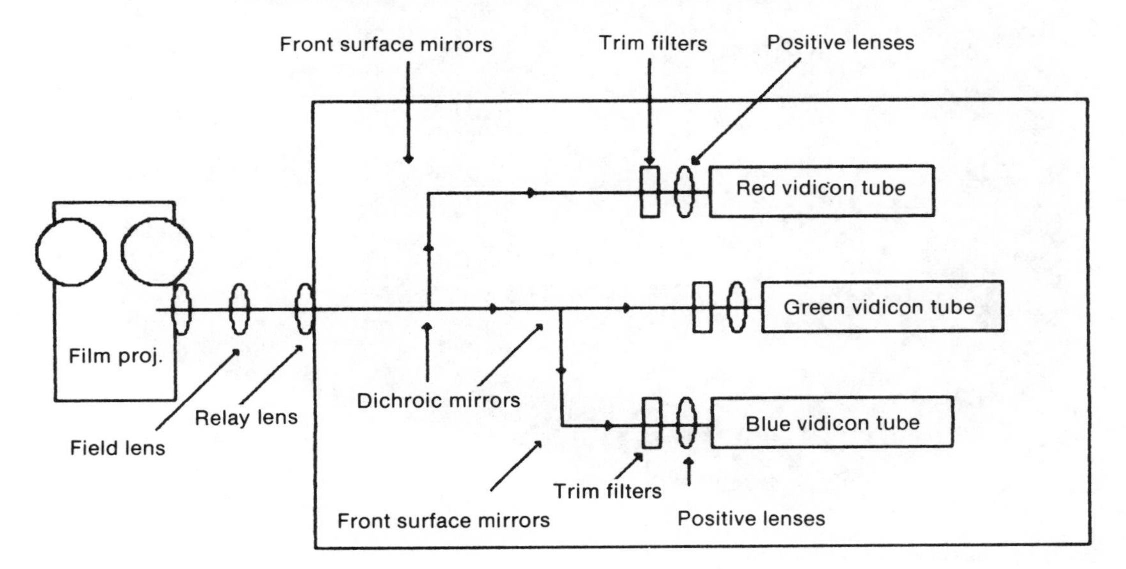

Figure 4.7: The MKIIIC Rank Cintel Flying Spot Scanner Telecine

Photo courtesy of Rank Cintel Ltd.

Figure 4.8: A Flying Spot Scanner Telecine Camera

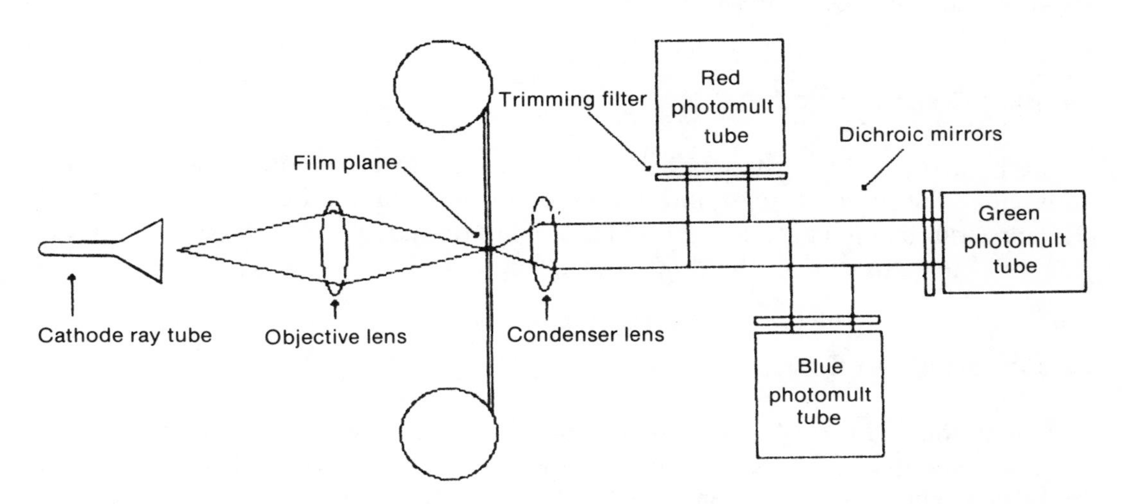

This allows the telecine operator to cue segments quickly without the risk of film damage.

X-Y Zoom

The X-Y zoom option enables the telecine operator to isolate specific areas of the film frame during the transfer process. The zoom range of a typical flying spot system is 6:1, with preset stops of 1.85:1 and 1.66:1 for wide-screen film formats.

Varispeed

The variable speed feature allows telecine operators to increase the film speed to 30 frames per second or decrease the speed to 16 frames per second. As discussed in Chapter 3, the "speed up" option is often used to reduce the running time of films that are being transferred for broadcast in TV time slots. Telecine systems that offer this feature must also offer connections for adding the pitch correction equipment necessary to ensure that the audio track still sounds normal after the transfer.

Cinemascope Compensation

As discussed in Chapter 3, television features a different, and much more restrictive, aspect ratio than film—particularly if the film was shot in cinemascope. If the telecine system features a cinemascope compensation option, it will automatically adjust for the TV aspect ratio during the transfer. This is done by either blanking the top and bottom of the picture to retain the original wide-screen effect, or by performing a "panscan" to enlarge selected sections of the picture area and establish the correct aspect ratio.

Primary Color Correction

On most telecine systems, the operator can use three remote joystick controls to balance the three primary and three secondary colors in the transferred image. This programmable color correction feature allows the operator to adjust both the hue and the saturation of the three primary colors.

Secondary Color Correction

With the addition of a variable matrix (an optional feature on many flying spot telecine systems), the operator can adjust the hue and saturation of the three primary and three secondary colors on a programmable, scene-by-scene basis.

Sound Reproduction

The flying spot scanner telecine system can reproduce sound from either optical or magnetic sound tracks on 16mm film, or from the optical track on 35mm film. The 35mm magnetic film tracks are transferred using the "double system" method. That is, the film is synchronized with an external magnetic film audio transfer machine.

LINE ARRAY TELECINE

The line array telecine is still another, very different type of telecine system. Unlike camera tube or flying spot scanner systems, the line array does not use a vidicon or any other electron tube–type image-sensing device. Instead, it employs all solid state circuitry. As a result, rather than use the normal interlaced scanning technique to form the TV signal, the line array system scans the image line by line as the film frame moves through the gate. The signal created by this system is converted to a standard, interlaced TV signal later in the process. (See Figure 4.9.)

Like the flying spot scanner system, the line array telecine can accommodate 35mm, 16mm or Super 8mm film in either positive or negative form. To change film type, the operator simply changes the film gate unit. In addition, like the flying spot scanner, the line array employs a capstan servo system for moving the film. This makes for fast and precise film movement, and it reduces the risk of damaging film during transfer.

The term "line array" comes from the method used to scan the film image during the transfer process. As shown in Figure 4.10, the film first passes through

Figure 4.9: The Marconi B3410 Line Array Telecine

Photo courtesy of Marconi Broadcasting Div., The Marconi Co. Ltd.

Figure 4.10: Line Array Optical Arrangement

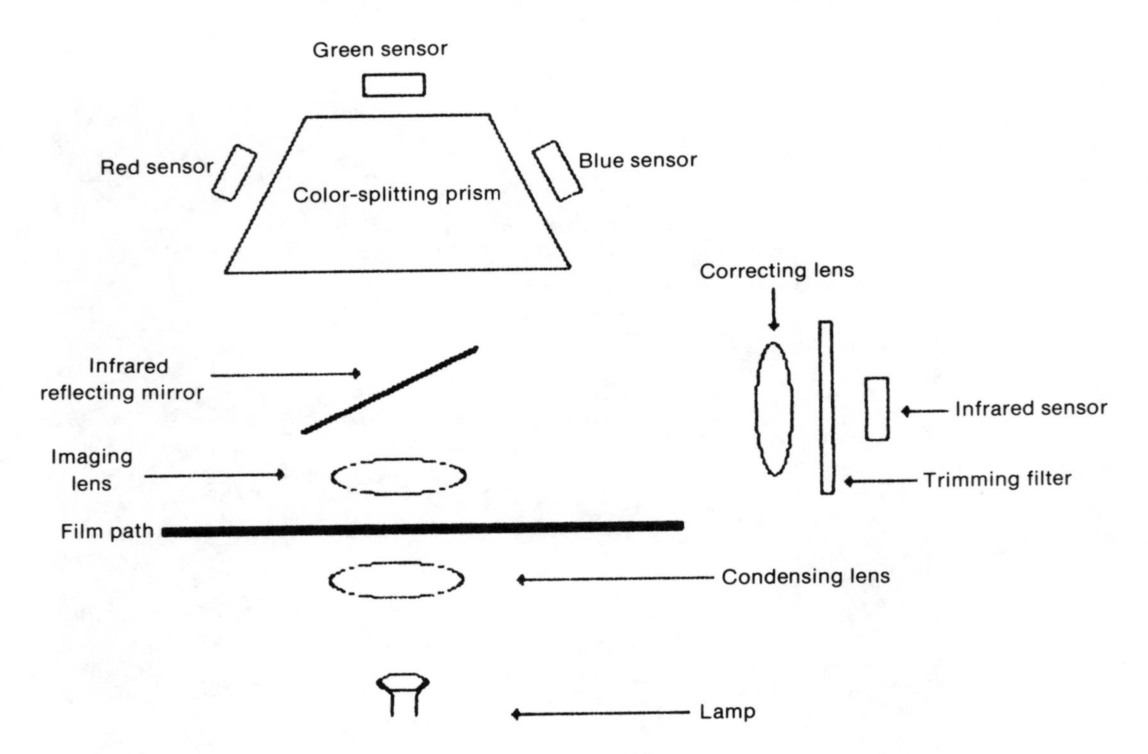

the line array system's picture gate, where it is illuminated by a 500-watt halogen DC lamp. The illuminated image is directed to the color-splitting prism, which separates the image into its red, green and blue components.

Next, the separated image is picked up by the system's solid state sensors. The line array system has three identical photosensitive image sensors, one for each primary color channel. Here's how the image sensors work.

As the film moves through the picture gate, the separated image is detected by the photosensitive diodes on a horizontal, line-by-line basis. During the horizontal blanking interval of the TV signal, the information detected through this process is transferred to two charge-coupled device (CCD) shift registers. This frees the photosensitive diodes to detect the next horizontal line of the image.

As the diodes turn their attention to detecting the next line, the CCD shift registers transfer the image information that they just received to the system's video circuitry. This charge/store/discharge process continues until an entire film frame is "scanned" sequentially, line by line.

Although this line-by-line process produces a high-quality video image, there are a few hitches. For example, because the vertical component of the transferred image is generated solely by the motion of the film through the line array system's frame gate, it is impossible to still-frame a picture during the transfer process—unless you are willing to add digital frame store devices to the system. This is also a problem during color correction, since the film has to keep moving during the transfer. To date, the only way to get around this problem is to use a slow film jogging function, a rather awkward feature that has met with less than enthusiastic response from telecine operators.

Despite these drawbacks, line array systems offer several advantages over other telecine systems. For example, the solid state components are highly reliable (the performance of CCD sensors doesn't suffer from the same "deterioration with age" problem that afflicts vidicon or cathode ray tubes), and the digital circuitry virtually eliminates registration problems and image lag. Also, like the flying spot scanner, line array systems operate quietly. Other features and advantages of the line array telecine system are described below.

Variable Speed Shutter Control

Taking advantage of its variable speed shutter control, the line array system can move film at 10 times (for 35mm film) and 25 times (for 16mm film) normal projection speed. This allows the telecine operator to cue segments quickly without the risk of film damage.

Slow Motion Forward or Reverse

The line array telecine can produce a broadcast-quality slow-motion picture at one-half the normal film speed. The slow motion can occur with the action running in forward or reverse.

Programmability

With the addition of the optional PREFIX programmable control system, the telecine operator can program any line array functions to operate automatically. The operator can also use PREFIX to enter location data for high-speed cueing. This information can then be stored on a computer disk and retrieved for later use.

Cinemascope Compensation

If the film being transferred was shot in cinemascope, the line array system can adjust the picture to fit within the TV aspect ratio. This is done by either blanking the top and bottom of the picture to retain the original wide-screen effect or by performing a "panscan" to enlarge selected sections of the picture area and establish the correct aspect ratio.

Sound Reproduction

The line array telecine can reproduce the sound from magnetic Super 8mm sound tracks, optical and magnetic 16mm tracks, and optical 35mm tracks.

Negative or Positive Film

The line array system can be used to transfer positive film or negative film.

TELECINE SYNCHRONIZATION SYSTEMS

One of the latest advances in electronic post-production technology is the telecine editing and synchronization system. The first of these systems, the automatic video replacement system (AVRS), was introduced by Calaway Engineering in 1980. The AVRS permits frame-accurate synchronization between the Rank telecine system and a 1-inch video recorder equipped with a CMX 340 interface—a device that allows the interconnection of and communication between two different pieces of equipment. Using this setup, telecine operators can replace existing segments on the videotape with film material sent directly from the telecine.

The newest incarnation of telecine synchronizing technology is the Time Logic Controller (TLC). Introduced by Time Logic Systems, Inc., at the 1986 convention of the National Association of Broadcasters, the TLC is an advanced controller capable of interconnecting one or two Rank telecine systems with any combination of up to four video or audio recorders.

The TLC operates much like an online video editing system, allowing the interconnection and control of the telecine system, audio and video recorders, a video effects switcher, and an audio mixing board. Significantly, the TLC can also control the general-purpose interfaces used to trigger and manage digital effects, ultimatte keying and other special effects. Features of the time logic controller include

- both NTSC and PAL operation
- both assembly and insert editing
- ability to perform cuts, dissolves, wipes, split edits and delay dissolves
- variable speed operation with automatic calculation of speeds
- edit point entry in SMPTE time code, feet and frames, seconds and frames or absolute film frames
- field-accurate edits

- capacity to work with any electronic device controlled through a serial interface
- ability to program the system to meet individual customer's needs
- disk storage of editing decisions

To the producer, all these features make the TLC a very flexible and powerful system that offers them full control over the film transfer process. Even more important, for projects shot on film, the TLC allows producers to create an edited master videotape directly from the film negative.

The TLC also offers some other important benefits. For example, using the TLC system, producers can create master syndication tapes in four separate formats, assuming that the system controls four video recorders. In addition, damaged or problematic shots on transferred masters can be corrected or replaced with one-field accuracy. This can be particularly useful when wide-screen features are being transferred for television, since the title sequences must often be redone to fit the TV aspect ratio and edited onto the master tape.

The TLC system also allows the transfer of dailies with the synchronized ¼-inch nagra sound track, eliminating the need to make a mag film copy. To do this, the video colorist (telecine operator) must use a proprietary syncing system developed by the Randken Corp. In addition, with the variable speed option available on the Rank telecine system, the video colorist can use the TLC to alter the length of a film so it fits within TV time slots.

As video colorists gain more experience with the TLC system, they are sure to uncover other applications for this innovative telecine system. In fact, using the TLC, it may even become commonplace for the producers of television commercials and other filmmakers working on TV projects to edit their material straight from the film negative to videotape.

IMAGE BALANCING AND THE VIDEO COLORIST

The video colorist is the telecine operator—the person responsible for setting up and running the various pieces of hardware that comprise modern telecine systems. But the video colorist is much more than simply an equipment operator. Along with running the telecine system, the colorist is responsible for achieving the final look and "feel" of the film negative or print as it is transferred to videotape. On some projects, this involves making rather minor adjustments to overall color or contrast characteristics. On other projects, it can involve extensive, frame-by-frame recorrection of the film image, or a "pan and scan" correction of a wide-screen feature so it will fit within the 4:3 television aspect ratio.

Today, the total responsibilities of the video colorist include

- film cleaning
- telecine operation
- color-balancing adjustments
- colorimetry adjustments
- pan and scan operations
- film-to-video editing
- transfer and mixing of stereo sound tracks

Film Cleaning

Before colorists load any film on the telecine equipment, they must make sure that the film is clean. This holds true regardless of whether the film is a negative fresh from the lab, where it should already have been cleaned, or a visibly dirty print that is in clear need of an ultrasonic "bath." A dirty print will result in a poor-quality transfer image, and dirt and dust can raise havoc with the critical tolerances of the transfer equipment.

Telecine Operation

With today's elaborate telecine systems, the video colorist must be skilled in the operation of many different pieces of highly sophisticated equipment. For example, along with knowing how to coax the best possible results from the film chain hardware, colorists must know how to work the mag playback machines used in double system transfer and the various digital effects options that are available on most modern telecine systems. Just as important, colorists must know how to operate color correction equipment.

Color Balancing

Before the transfer session can begin, the video colorist must "color-balance" the film. First, the colorist uses a standard gray scale to adjust the primary color components and contrast values. Then, the film itself is loaded on the telecine, and the colorist previews and evaluates a sample segment.

Through this evaluation process, the video colorist determines the type and degree of color correction required for the film. Usually, films fall into one of three color correction categories:

- those that require no color correction
- those that require some overall rebalancing of the contrast values and primary colors
- those that require rebalancing of the contrast values and primary colors *and* readjustment of the complementary colors (cyan, magenta and yellow)

As they prepare for the transfer, the client and video colorist will usually work together to establish the look and "feel" of the film. First, they will use the black, gamma and luminance controls to adjust the dark areas, midrange grays and bright areas of the film image. Then, they will use the individual color controls to adjust the primary and (on some films) complementary colors. In the color-balancing stage, these parameters are usually programmed into the color corrector memory and used throughout the transfer.

Colorimetry Adjustments

The colorimetry process is a more specific and sophisticated version of the color-balancing process. In color balancing, the adjustments made to color and contrast values are usually applied to the entire length of the film. In colorimetry, the adjustments are made on a scene-by-scene or frame-by-frame basis, and the overall picture balance is not affected.

Usually, the video colorist will enter the colorimetry corrections into the telecine system's memory as a sequence of frame numbers and color parameters. During the transfer, the system will call up these command sequences and automatically perform the appropriate corrections.

Video colorists often use the colorimetry process to fine-tune a specific shot or scene. For example, if the film being transferred is footage from the production of a TV commercial, the colorist might decide to enhance only the colors in the product itself, leaving the colors in the surrounding footage set at their current levels.

In addition, colorimetry often plays an important role in the transfer of footage shot as part of rock video productions. Often, the footage for rock videos is shot in several different locations under widely, and wildly, varying production conditions. When this is the case, colorimetry adjustments can help make the separate segments seem more closely matched.

Pan and Scan

As mentioned earlier, a film shot in cinemascope cannot be transferred to video without making some adjustments for television's more restrictive aspect ratio. Making these adjustments is the responsibility of the video colorist, usually in consultation with the film's producer or director.

Currently, there are two ways to resolve the "wide-screen-film-to-small-screen-television" transfer problem. First, you can do what is often done in Europe—establish borders at the top and bottom of the TV screen, so the film image appears as a band in the middle of the screen. Although this technique retains the original film aspect ratio, it cuts off a good percentage of the TV screen. According to many producers, it also looks "funny" to many members of the American TV audience. As a result, this solution has never really caught on in the United States.

The second solution, and the approach most often used in the United States, is the "pan-and-scan" technique. In the pan and scan, the telecine system is programmed to focus on only one portion of the wide-screen feature at any one time. The system can be told to switch instantly to another section of the wide-screen image, or to move (pan) across the image to establish a new perspective.

If all of the primary action will not fit on the TV screen, the producer or director must decide which portions of the action to show at any one time. For example, if both characters in a wide-screen dialog scene can't be squeezed onto the TV screen at the same time, the producer or director might choose to use the pan-and-scan technique to rework the scene into a series of single shots that alternate between the two characters. Or, if there is not a great deal of rapid-fire interchange between the characters, the producer might decide to have the telecine pan from one character to the other, in a motion very similar to the conventional camera pan used in film and television production.

Film-to-Video Editing

Using systems such as the TLC described earlier, video colorists can now perform frame-accurate editing during the transfer process. Colorists can use this editing capability to replace shots or scenes on a transfer tape with new film material and to perform more sophisticated sorts of edits and special effects. In addition, as I discussed earlier in this chapter, colorists may soon be using the TLC and similar systems to derive an entire edited master tape directly from the film source material.

Stereo Audio and Mixing

To accommodate projects that require stereo or multitrack sound work during film-to-tape transfer, many modern telecine suites are equipped with multichannel audio mixing boards. On most productions, the video colorist must coordinate this "sound transfer" along with the film transfer.

Usually, the sound tracks are first recorded onto multitrack audiotape. Then, during the transfer session, the tracks are passed through the mixing board, synchronized with the film and recorded on the master videotape.

CONCLUSION

At the beginning of this chapter, I defined two types of film material used in television. First, there are films that were originally completed as theatrical productions but that are now scheduled for TV or videotape distribution. Transferring these film productions is usually a fairly straightforward job involving only modest color correction.

The second category of film material—film footage produced specifically for television—can present more of a challenge. The film segments are often shot at a variety of locations under widely varying conditions, and they must be intercut with footage from film libraries and other sources. When this is the case, it's up to the video colorist to pull the segments together, performing whatever corrections and adjustments are necessary to make sure that the transferred images "match up" on the master videotape.

As this description suggests, the video colorist has become a critical member of the post-production team. Along with knowing how to work some highly sophisticated equipment, the colorist must know how to work with production personnel to achieve the proper look and feel in the transferred film. In addition, with the advent of telecine synchronization systems, video colorists must know how and when to add video transitions and special effects.

In the post-production telecine process, the final product is usually a videotape transfer reel along with a videotape workprint ready for offline video editing. At this point, once the videotape workprint is checked and approved, the video colorist's job is over. However, for several members of the electronic post-production crew, the work is just beginning.

5 The Offline Editing Process

The offline editing stage of electronic post-production is analogous to the workprint stage of conventional film editing. In both processes, the goal is the same: to produce an edited workprint copy of the program and to obtain an edit decision list for use in final conformation. Of course, in conventional offline video editing, the work is all done electronically. That is, the videotapes containing the workprint copies of the transferred production footage are never physically touched or spliced. Instead, the transferred scenes and segments are played back on one videotape machine (the playback VTR) and recorded on a second one (the record VTR), usually one segment at a time. As a result, the original transfer material remains safe and intact on the source reel, ready to be used in the final online conformation stage.

As mentioned above, the offline editing process produces two "products": an edited workprint and a corresponding edit decision list. Copies of the videotape workprint are then sent to producers, network executives or company officials for final approval before committing the project to online conformation, when the original transfer tapes and a revised and "cleaned" editing list will be used to assemble the finished program.

On feature films or other productions that are conformed in the conventional film manner, offline editing can still prove valuable. For these projects, the primary purpose of offline editing is to produce an edit decision or negative cutting list that will guide the slicing and splicing of the film negative in the final conformation stage. For more information, see the discussion of feature film post-production in Chapter 2.

As this brief discussion suggests, the edit decision list is the key component of the post-production process. How this list is created depends on what type of editing system you are using. On simple and inexpensive offline editing systems, editors must generate the decision list themselves by "hand logging" the edit points used in piecing together the workprint. On more expensive systems, a computer does the logging for the editor, generating an edit decision list on a paper printout, paper punch tape or, increasingly, floppy disks (see Appendix A).

In this chapter, I describe the strengths and limitations of several types of offline editing systems, analyzing how the offline process differs depending on the capabilities of the system. First, to set some parameters, I'll describe the features of the ideal offline editing system, as that dream system might be envisioned by several experienced film and video editors.

THE IDEAL OFFLINE VIDEO EDITING SYSTEM

Suppose several film and video editors were given free rein to develop the offline system of their dreams. What features and functions would they want? Speed and accuracy, of course, but what else? Here are some features that would probably make their "most wanted" list:

- A computerized editing control console that allows editors to enter frame-accurate editing points quickly and easily, using as few keystrokes and controls as possible. The console should also allow editors to program any "out of the norm" features as soft-key functions.*
- The capacity to modify any picture or sound element and to position or reposition elements in relation to each other.
- A full random-access capability that allows editors to preview and record individual edits or large groups of edits.
- A feature that registers time code and edge numbers and automatically compiles the edit decision list, leaving post-production personnel free to concentrate on making creative editing decisions.
- Simple and fast procedures for entering log sheet data, with provisions for finding and reading that data quickly and accurately.
- The ability to memorize and store each editing element and to recall and change any element or group of elements with a few simple keystrokes.
- A feature that allows post-production personnel to open the edit decision list and to shift, insert or delete elements. This feature should also include a "ripple" function that automatically adjusts the full list to accommodate the changes.
- The ability to control a multitrack recorder or group of multitrack recorders, or to coordinate multitrack digital audio processing.
- The capacity to generate edit decision lists in sound only, picture only, and sound and picture form. The lists should include both time code and key codes, and it should be accurate within one TV field.

---

*Soft keys are keys on a computerized control console that the editor can program to perform a preset function or series of functions.

- Some capacity for performing fades, dissolves and other simple sound and video special effects.
- The ability to generate a high-quality edited workprint with broadcast-quality multitrack sound.

At this time, no offline editing system offers all of these functions and features. However, if advances in computerized editing and digital video technology continue at their present pace, this "dream system" will soon become a reality. Until then, editors will have to settle for using the features found on one of the three types of offline systems that are currently available: control track systems, automatic time code systems and computerized systems.

CONTROL-TRACK OFFLINE EDITING SYSTEMS

If you consider only per-hour equipment costs, control-track systems are by far the "best buy" in offline editing. In 1986, the weekly rental rate for a simple, two-VTR, cuts-only control-track system was approximately $700 to $1000. Pretty reasonable, considering that this allows an editor to use the system 24 hours a day for seven full days. Monthly rates are also available, usually at a lower per-day average.

But equipment rental fees are not the full story. Although control-track systems are relatively inexpensive, they require the editor to cue each source tape manually to the proper editing points and to log all edit decisions by hand. As a result, for those projects that will move on to time code–based video conformation, control-track offline editing can end up costing more money than it saves. In particular, there is the cost of the extra time and staff required to enter the hand-logged edit numbers into the online system's computer memory, as well as the risk of human error that this introduces to the editing process. To alleviate this shortcoming, some newer upgraded versions of control-track edit systems, such as the one shown in Figure 5.1, are now available with the ability to read time code and print an EDL.

AUTOMATIC TIME CODE OFFLINE EDITING SYSTEMS

Automatic editing using time code equipment, like that shown in Figure 5.2, is usually a faster and more efficient process than control-track offline editing. Depending on the sophistication of the equipment, automatic time code offline editing can offer the following advantages:

- the ability to read longitudinal and vertical interval time code or control-track pulses
- the ability to enter edit points in time code, so several playback VTRs can be automatically cued to the correct locations

Figure 5.1: The Super90 Control-Track Edit System, an Upgraded Version of the ECS-90 System

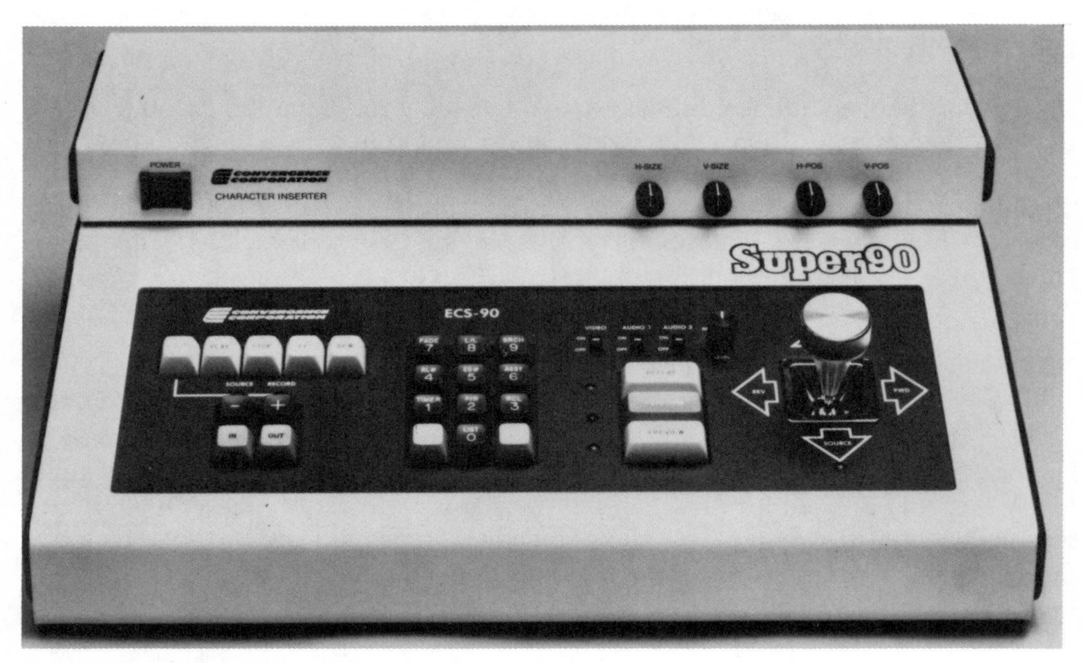

Photo courtesy of Convergence Corp.

- automatic preparation of edit decision lists in computer-readable form (punch tape or floppy disks)
- automatic printing of edit decision lists
- automatic control of video switchers and audio mixing boards

In short, compared to control-track systems, time code editing systems allow editors to perform more offline chores automatically, so the editor can spend more time concentrating on the aesthetics of the project. In addition, time code systems make it much easier for editors to prepare accurate, computer-readable edit decision lists—an essential item for successful online conformation.

COMPUTERIZED OFFLINE EDITING SYSTEMS

Essentially, most conventional computer-controlled systems used in offline editing are identical to their computerized counterparts used in online conformation. The only real differences are that offline systems are capable of fewer video effects and that they use small-format VTRs.

Let's take the GVG (Grass Valley Group, Inc.) system, the control keyboard for which is shown in Figure 5.3, as one example of a computerized offline configuration. The GVG can function as either an offline or online (conformation) editing system. The controller shown in Figure 5.3 can coordinate up to seven playback and record VTRs, in any combination. The VTRs can be ½-inch Betacam or ¾-inch U-matic cassette machines, or 1-inch or 2-inch tape machines. Using the keyboard, editors can also control multitrack audio recorders, audio console faders and video switchers.

The GVG system features most of the conventional editing capabilities, including automatic assembly of edits, audio and video "split" editing, variable speed motion and edit list management. It also offers a film list entry mode that converts film footage counts entered into the system to TV time code.

When all the editing is done, the GVG produces either a finished workprint or a final master videotape, depending on whether you are using the system in offline editing or final online conformation. It also produces an edit decision list as a hardcopy printout, or on paper tape or floppy disk. At this time, however, the GVG will not produce a negative cutting list.

RECENT DEVELOPMENTS IN COMPUTERIZED OFFLINE SYSTEMS

Any new area of product development will invariably result in many successes and failures. I have therefore limited my discussion to just three notable examples of new technology: the Montage Picture Processor, the EditDroid system and the SoundDroid system. Several other products will undoubtedly come on the market in a very short time, but these three examples will help the reader understand in which direction the film and television industries seem to be heading.

Figure 5.2: The Sony BVE-900 Time Code Video Editing System

Photo courtesy of Sony Broadcast Products Div.

Figure 5.3: Computerized Video Edit Controller

Photo courtesy of The Grass Valley Group, Inc.

Montage Picture Processor

Introduced at the 1984 convention of the National Association of Broadcasters, the Montage Picture Processor is designed to function as an easy-to-operate, random-access editing system capable of handling either single-camera film-style productions or multiple-camera video-style projects (see Figure 5.4). The Montage system can accommodate editing data for as many as 2500 separate scenes. In addition, with the extended memory module option and a sufficent supply of playback VTRs, the Montage Picture Processor allows editors to preview up to one hour's worth of program edit decisions.

Advantages of the Montage Picture Processor

Compared to other computerized editing configurations, the Montage system features a much simpler control panel and much more streamlined editing pro-

Figure 5.4: Montage Picture Processor

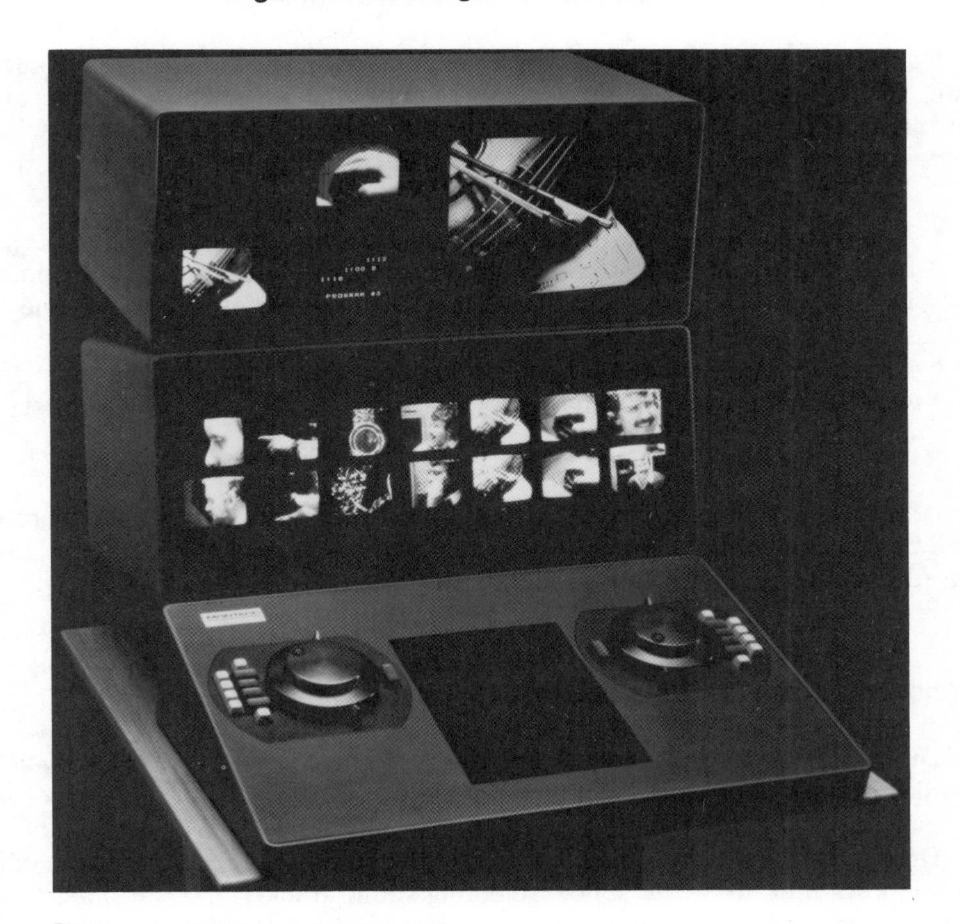

Photo courtesy of Montage Computer Corp.

cedures. In addition, by taking advantage of the Montage system's extended preview function, post-production personnel can preview and evaluate entire sequences of edits without actually performing the edits. This allows editors to trim, change or delete edit points quickly and easily, without wasting time recording test edits on videotape. On many projects, the post-production crew will not bother committing any edited material to videotape during the offline session, unless the producer has requested a workprint for screening.

The Montage Offline Process

Before the Montage offline process can begin, the dailies from the production session must be transferred to 1-inch videotape and ¾-inch videocassettes. This material is in turn transferred to either 7 or 14 ½-inch Beta-format VCRs, depending on how many are connected to the system. All of these "extra" VCRs are necessary to permit the Montage controller to find sequences from the production quickly and accurately. While one Beta VCR is searching for the next scene to be previewed, the others can be searching for subsequent scenes.

During the transfer to VCR, an assistant editor creates an "electronic log sheet" by typing scene and take numbers from the production footage at the controller keyboard. Once the editor has previewed all of the edits and selected edit points, the Montage system will provide one or all of the following:

- A negative cutting list, including a printout that shows edge numbers and frames, feet and frames, total numbers of frames and sound code numbers. In addition, the printout will indicate the location of any two-frame "hot splice" proximity edits.
- A conventional video edit decision list on a standard 8-inch floppy disk.
- A workprint copy automatically recorded on ½-inch or ¾-inch videocassette.

Needless to say, the negative cutting list feature will be used to conform the project in the conventional film manner, while the time code–based edit decision list will be used to conform the project on videotape.

Components of the Montage System

The basic Montage Picture Processor System consists of a workstation, a central processing unit and a memory drive unit.

The workstation is designed to simplify the editing process. All of the editing function keys (cut, dissolve, splice, selecting sound track 1 or 2, etc.) are located on the editor's left side, and all of the processing keys (trim, pull, discard, play, insert, etc.) are located on the editor's right side.

The picture-processing knobs and levers used for calling up work bins, accepting edit decision changes and scrolling through picture displays are also situated on either side of the editor. In between, there is an electronic blackboard, called the "digitizing pad," used for entering notes with an optional electronic stylus (light pen).

The monitor display includes 14 3.7-inch monochrome monitors that display the heads (beginnings) and tails (ends) of seven different shots. The head and tail of the individual shot selected for preview or editing are displayed on two 5-inch monochrome monitors, with the random-access preview mode displayed on a 13-inch color monitor.

At the heart of the system is the central processing unit (CPU). In the Montage Picture Processor, the CPU is a completely self-contained turnkey computer system with a "power failure recovery" feature that protects data should a power outage occur. The Montage CPU also features proprietary picture and sound label digitizers and smooth-scrolling picture label display processors, which are displayed on the 14 3.7-inch monochrome monitors.

The memory drive unit includes seven Sony Beta hi-fi stereo transports with individual interfaces and routing systems, an internal video switcher for dissolves and wipes and a sound mixer for level adjustments and fades.

The Montage System has several optional items, including a memory unit for an additional seven VCRs, an "electronic grease pencil" (light pen) and a storyboard software feature. If the optional printer is attached to the system, the storyboard software allows post-production personnel to print storyboards that contain picture and sound editing data for each scene.

The EditDroid

Introduced by Lucasfilm Ltd. at the 1984 convention of the National Association of Broadcasters, the EditDroid system is a random-access editor designed to reduce or eliminate much of the tedium and frustration associated with the mechanics of editing.

When used with source material recorded on laser disc, the EditDroid system offers rapid, random access to shots and scenes, as well as unlimited preview capabilities (entire scenes can be previewed without actually being recorded). Editorial decisions can be made, reviewed and refined in real time, in a process that requires no lab reprints and that results in no generational loss of image quality. The EditDroid is designed to work with any combination of 1-inch, ¾-inch and ½-inch videotape recorders, as well as with laser video discs.

As shown in Figure 5.5, the EditDroid's workstation includes a source screen on the left for reviewing dailies, an edit screen in the middle for previewing edit

Figure 5.5: The EditDroid

Photo courtesy of Lucasfilm Ltd. EditDroid is a trademark of Lucasfilm Ltd.

sequences and an electronic logbook on the right for viewing log sheets and edit lists. Other components include the touchpad used to control the EditDroid and a keyboard for typing log notes and other text into the system.

The EditDroid can generate either a conventional editing list for online editing on videotape or a negative cutting list for conforming the project on 16mm, 35mm or 70mm film.

The SoundDroid

Currently under development at Lucasfilm Ltd. as the companion to EditDroid, SoundDroid is being designed to organize, simplify and automate the audio post-production process. The SoundDroid system will function as a single, general purpose workstation that can be used to perform sound recording, editing, mixing and equalizing—functions that require separate components on most conventional audio systems.

The SoundDroid system will feature all of the sound processing capabilities of a conventional mixing board. In addition, because its circuitry will be completely digital, the SoundDroid system will store, record, edit, mix and process sound information, as well as reproduce sound instantly upon the operator's command. Also, because the sound tracks will be stored digitally on computer disks, SoundDroid will allow post-production personnel to slip, edit and reedit tracks quickly and accurately.

As shown in Figure 5.6, SoundDroid's workstation will include a high-resolution, touch-sensitive computer screen. The screen will display various menus of options, depending on the type of sound work being performed. Below the touch screen there will be eight motorized fader controls that, along with the knobs to the right of the screen, can be assigned to perform different tasks. The final system component, a video display monitor, will be located directly in front of the system operator.

CONCLUSION

Offline editing is becoming an increasingly sophisticated process. Although the "dream system" described at the beginning of the chapter is not yet a reality, recent developments in computerized editing systems are making offline editing a much more flexible and precise process. In fact, for projects that will be conformed

Figure 5.6: The SoundDroid

Photo courtesy of Lucasfilm Ltd. SoundDroid is a trademark of Lucasfilm Ltd.

on videotape, the boundaries between offline and online editing are becoming increasingly blurred, with the same computerized systems capable of functioning in either capacity.

Despite all of these technical developments, the goal of offline editing remains quite simple: to produce an edited workprint copy of the program and to obtain an edit decision list for use in final, online conformation. Once that edit decision list has been generated, and once the workprint copy to which the list corresponds has been approved by production executives, you are ready for the final stage of electronic post-production—the online editing process.

6 Video Conformation: The Online Editing Process

You have finished the offline editing session, and the videotape workprint has been approved by production executives. In addition, any changes requested by the executives have been incorporated in the edit decision list. Now, if the production is to be finished on videotape, you are ready for the final stage of electronic post-production: online video conformation.

Video conformation, or "online editing," is the stage of electronic post-production in which material selected from the original transfer reels is pieced together to form the final, edited master reel. In conventional film editing, this is analogous to cutting the negative, finishing the opticals, correcting color timing, performing basic sound editing and mixing, and creating final release prints. The difference in electronic post-production is that all of these functions are usually performed in one session. As a result, online conformation is the most technically demanding stage of electronic post-production.

Before beginning online conformation, you must have one of the following: an accurate film count list for conversion to time code equivalents; a hand-logged, time code–based edit decision list; or a computerized edit decision list stored on punch tape or floppy disk. Generating the edit decision list on punch tape or floppy disk is the preferred option, since these formats allow editors to enter the information into the editing computer quickly and accurately.

THE ONLINE EDITING BAY

Online editing facilities come in a variety of shapes and sizes, from single rooms equipped with inexpensive ¾-inch editing systems to posh suites outfitted with the latest in computerized editing technology. As Figure 6.1 shows, minimum

Figure 6.1: Minimum Equipment Configuration for an Online Editing Bay

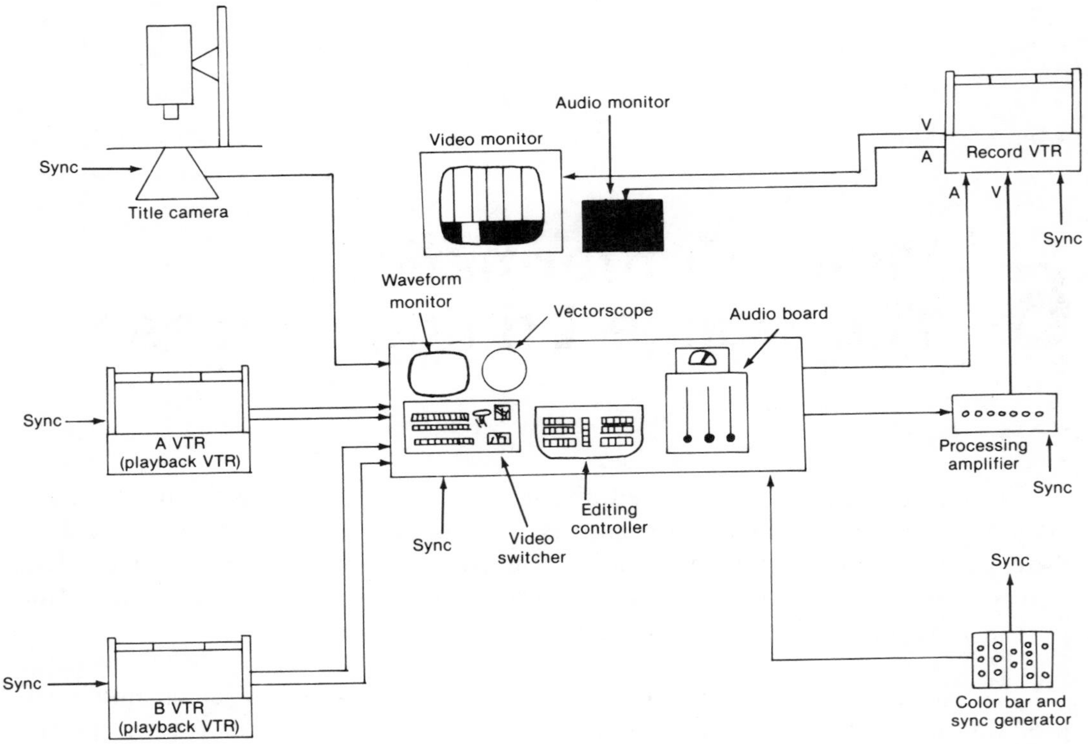

equipment requirements for a professional-quality online editing bay include the following:

- an editing controller
- two playback VTRs with time base correctors
- one record VTR capable of assemble and insert (audio/video, audio-only and video-only) edits
- a video switcher with at least one special effects row
- various signal monitoring and processing components, including a waveform monitor, vectorscope display and video processing amplifier
- a master color bar and sync generator
- a black-and-white title camera
- a small audio mixing board with equalization capability
- audio and video monitors

In most modern online bays, the editing controller is a computerized component that is capable of "reading" an edit decision list from punch tape or a floppy disk. In addition, most modern bays are equipped with several additional playback VTRs, to allow for fast and accurate automatic assembly.

Once the edit decision list has been read into the editing controller, and once the appropriate transfer tapes are loaded on the playback VTRs, you are ready for automatic assembly of the edited master tape.

AUTOMATIC ASSEMBLY

Automatic assembly is the automated, computer-controlled process by which shots and scenes from the original transfer reels are pieced together (assembled) to form the final edited master reel. For the process to work efficiently, a "clean" edit decision list must be loaded into the editing computer, and the proper transfer reels must be loaded on the playback VTRs.

There are two different types of auto assembly: A mode (also called "linear" or "sequential"), and B mode (also called "optimum" or "checkerboard"). To choose between the two, you must understand how they differ.

The simplest method of automatic assembly is A mode. First, the edit decision list is loaded into the computer's memory, and the proper source reels for the first edits are loaded onto the playback VTRs. Then the editor gives the signal for the online system to start assembling. The system will continue to assemble the project in sequential order until it encounters an edit that requires a new source reel (a reel that is not currently loaded onto one of the playback VTRs). At this point, the computer will halt the assembly and ask for the new reel by its assigned reel number.

The more efficient form of auto assembly is B mode—as long as it is set up properly. In B mode assembly, the system performs *every* edit on the edit decision list that can be completed using the source reels that are currently loaded on the playback VTRs (with the exception of transition edits in which only one of the required reels is loaded). In other words, once the system has searched for and performed every edit that requires material from, say, source tape #3, you will be able to remove tape #3 from the playback VTR, stow it away for the duration of the online session and replace it with a new reel. As a result, B mode assembly greatly reduces the amount of online time needed for changing source reels—a time saving that translates directly into a saving on online rental fees. However, to use this mode of assembly, the following conditions must prevail:

- The edit decision list must have no edit time overlaps.
- All time and length factors pertaining to the project must be final before online assembly begins.
- Color setup of production material should be properly prematched and not readjusted during assembly.

For any projects that comply with these conditions, B mode assembly is usually the preferred approach, with the possible exception of projects that require numerous video special effects or multiple-reel transitions. Generally, that type of project can be edited more efficiently using A mode.

One final note on auto assembly modes. Many computerized editing systems feature a function called "lookahead." On systems that offer this function, the editing computer scans ahead on the edit decision list and pre-cues source reels to the proper location for upcoming edits. For longer editing projects, particularly those that will be assembled using B mode, this feature can save considerable time and expense.

ADDING TITLES AND GRAPHICS IN ONLINE CONFORMATION

The process of adding titles to an edited master videotape is surrounded by a variety of myths and misconceptions. For example, it's not uncommon for an advertising executive who is involved in the production of a TV commercial to show up at a video post-production facility with a title card in hand, naively expecting that he can simply add the titles to a scene that is already recorded on the edited master tape. This misconception stems from the quite common, but quite erroneous, belief that titles are merely "burned into" a previously recorded scene.

The Key Process

Our advertising executive would do well to memorize one of the fundamental precepts of videotape post-production. When a VTR is placed in the record mode, all video information that was previously recorded on that section of the tape will be erased, to be replaced by the video signal that is currently being fed into the VTR. For example, our executive might try adding a title card to an existing videotape scene by feeding the signal from a graphics camera into a VTR, cueing the videotape to the appropriate scene and pushing the record button. However, upon playing back the tape, the executive will find that there now exists a nice picture of a title card but no background scene, since that was erased during the recording process.

What the executive really wants to do is to "key" the titles over the existing scene. Keying involves using a video switcher to combine the background scene (cued on a playback VTR) with the title card image (as generated by a graphics camera that is focused on the card). As shown in Figure 6.2, the signals from the playback VTR and the graphics camera are sent to the video switcher, where they are combined into a single composite video signal that is fed to a record VTR. Remember, though, that the keying is actually performed at the video switcher/ special effects generator, with the record VTR simply standing by to accept the composite video signal fed from the switcher.

Figure 6.2: The Videotape Keying Process

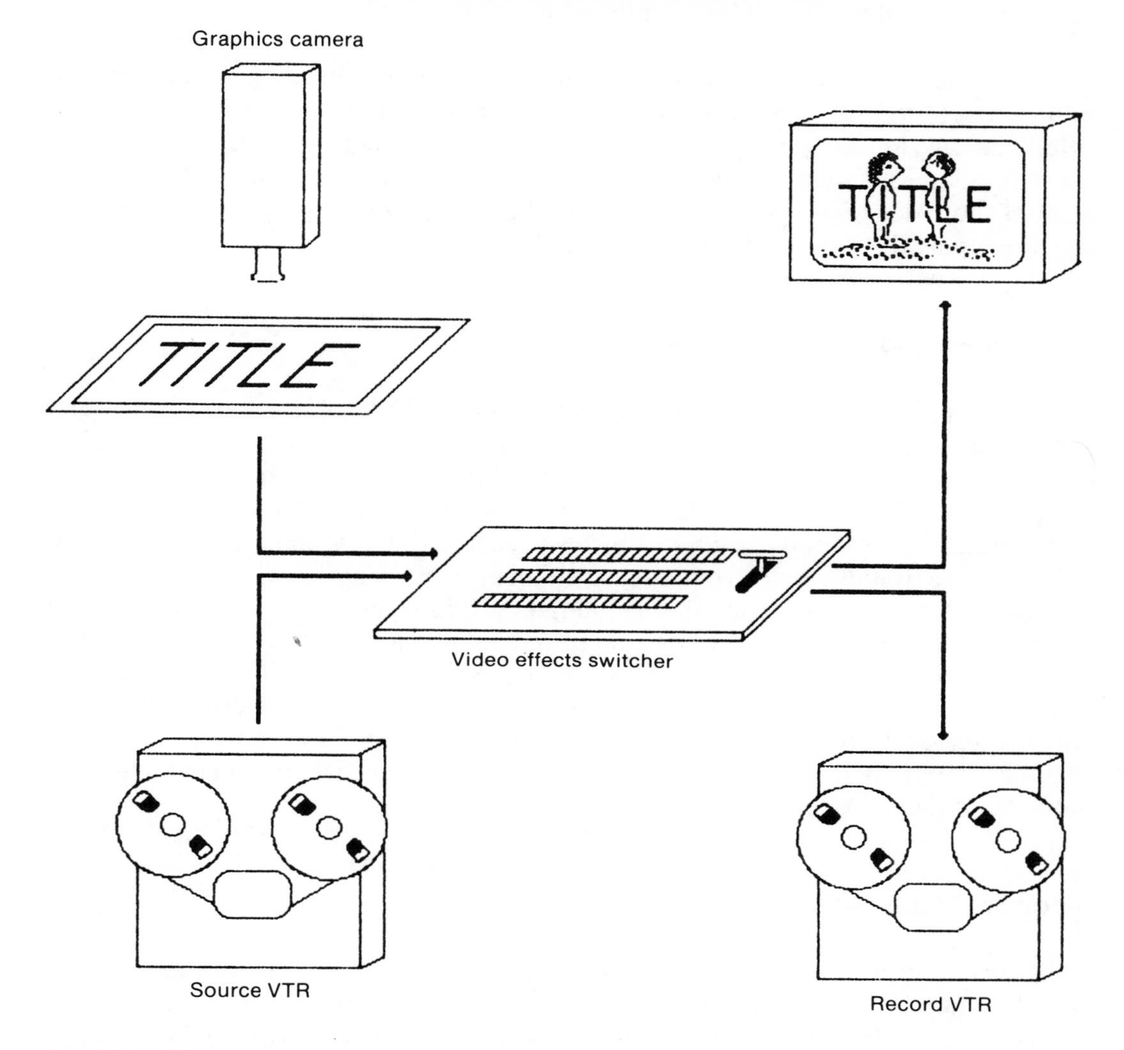

Graphics Techniques

All graphics and titles, whether generated electronically or prepared on title cards, should be approved and set up before the online assembly session begins. Nothing wastes more time and money than searching for and making decisions about graphics during the online process. Obviously, there are times when these decisions can be made only after viewing the edited scenes, as is often the case when the scenes involve digital special effects. As a general rule, however, titles and graphics should be prepared and approved well ahead of the final assembly session.

As explained in Chapter 3, all graphics must fit within the readable confines of the 3:4 television aspect ratio. In addition, TV titles should not contain fine-line detail in or around the letters. Finely detailed lines will not normally key properly, causing rough edges and "tears" in the video signal.

Graphics drawn on art cards should usually be prepared using white letters on a black background. If the graphic is drawn as black on white, make sure that the title cameras at the post-production facility can be electronically reversed to accommodate the change. Because the video switcher uses the highest luminance level of the video signal for its key source, the part of the graphic that is supposed to show up as the key must contain the higher luminance level.

Chroma Keying

In the process called "chroma keying" (also known to filmmakers as the "blue screen" process), color and luminance information are used as the key signal. The chroma key process takes a uniformly colored background as the key source and then replaces that colored background with the video signal from another scene. Any primary or secondary color can be used for the key signal, as long as the lighting is extremely flat, producing a uniform background without texturing or shadows. However, the image that is being keyed into the new background should not contain any of the key source background color, since that portion of the image will disappear along with the background. For example, a performer placed against a red key source background should not wear a red jacket. If he does, the jacket will mysteriously disappear when the key is performed. Blue is by far the most popular choice as a chroma key background source, primarily because skin pigmentation does not contain blue.

Electronic Graphics

Since their introduction in the 1970s, electronic graphics systems have experienced tremendous growth. Born as simple character generators capable of creating letters in a single font, electronic graphics equipment has evolved into today's sophisticated systems that offer several font styles, multiple colors and animation features.

Compared to art cards, electronic graphics were at first thought to appear cheap and artificial. However, as the technology has improved, electronic graphics have become widely accepted as the preferred option. In contrast to art cards, electronic graphics are fast and easy to change, and they provide perfect leveling and centering parameters.

Systems by Chyron Corp., Telemation, Inc., Aurora Systems, Dubner Computer Systems, Inc. and MCI/Quantel represent the current state of the art in graphics generators and animators. Most of these systems are capable of generating both single- and multiple-line titles, as well as title crawls, in white, colored and bordered font styles. Some of the systems also feature the ability to still-frame the video, to "paint" the frame and to draw animated sequences—and it's all done electronically.

IMAGE CORRECTION DURING ONLINE CONFORMATION

Once online automatic assembly is complete, how much image and color correction will the edited master require? That depends on a number of factors, particularly the conditions that prevailed during the production session. For instance, if the entire project was filmed in a single, controlled environment, such as a sound stage, very little color correction may be required. However, if the project was a multiple camera production shot at multiple locations under varying light conditions, much more image correction may be necessary.

Most producers of television commercials and music videos prefer to correct the image when the production negative is initially transferred to videotape. Because the amount of footage transferred for these productions is relatively low (typically less than 5000 feet), it is usually more cost-effective to correct the image during transfer than during or after the online conformation stage.

Admittedly, the telecine stage is not always the most appropriate or cost-effective time to perform color correction. For instance, correcting the image during the transfer session means that you will need to pay someone connected with the production to supervise the session, and this can add considerable cost. Often, this also means that you must move the transfer session to high-cost peak hours, rather than allowing the transfer to take place unsupervised during less expensive off-peak time. Of course, any money saved by not correcting the image during the telecine stage must be balanced by the extra cost and time spent correcting the image during or following online conformation.

Keep in mind, also, that many types of projects are impossible or very difficult to correct before they enter the editing stage. For example, a recent Linda Rondstadt concert shot for cable television used 11 cameras and required 250,000 feet of film. The production footage was transferred to videotape using only one light foundation (setting), and all image correction was performed after online conformation was complete. In this case, it would have been impractical to perform the image correction during the film-to-tape stage, particularly considering the amount of production footage involved.

Minor Image Correction During the Online Session

For projects that require relatively minor image correction during the online session, you have two choices. The first, and simplest, option involves manually adjusting the luminance, saturation and hue controls on the playback VTRs to achieve the look that you want. The second, more elaborate method requires passing the source material through some sort of color correction system. Using the system's controls, the image can be altered by adjusting the primary and secondary colors in the darkest, midrange and brightest areas of the image.

Unfortunately, because they are manual techniques, neither of these methods allows you to program the same set of changes for a series of shots. I recommend that you set the controls for a particular adjustment, correct *all* shots that need this adjustment, and then reset the controls and move on to the next series of shots. However, I do not recommend following this process on projects that require major color correction, since the costs associated with correcting each shot manually would be prohibitively high.

Major Image Correction After the Online Session

Large-scale image correction after online conformation requires special equipment and special expertise. Currently, I am aware of only one facility that is equipped to do the job quickly and relatively economically: Image Transform, Inc. in Burbank, CA. Image Transform has developed a proprietary process that permits programmable, frame-by-frame or scene-by-scene image correction.

The Image Transform process features many of the same functions offered by modern telecine systems (see Chapter 4). Along with permitting the rebalancing and correction of color and contrast values, the system features a video noise reduction option.

In the correction process, the signal from the original edited master tape is played through the system, while one or more copies of the corrected signal are recorded on the facility's VTRs. The result is a master tape that is cleaner than the original edited master tape, with no apparent generational loss.

At this point, let me add a general caution about video noise reduction. Although it can be a very valuable feature when used properly, more noise reduction is not necessarily better. Too much noise reduction can result in a "digitizing" effect—annoying vertical disturbances in the darker areas of the image. If only minor color correction is necessary, I recommend performing any noise reduction that might be required during the initial film-to-video transfer session. Unless you are performing color correction in the online session, there is no point taking the edited master down a generation simply to accommodate noise reduction.

ELECTRONIC SOUND POST-PRODUCTION

Electronic sound post-production, known in video as "audio sweetening," has made great strides since the advent of electronic time code. Using conventional audio sweetening techniques or the more sophisticated post audio processing (PAP) system developed by Glen Glenn Sound, producers of projects shot on film now have viable electronic alternatives to the mechanical methods used in conventional sound post-production. In other words, using video equipment and multitrack audio recorders, it is now possible to perform high-quality sound editing as part of the electronic post-production process.

Electronic sound post-production usually includes five distinct stages: layover; spotting; sound effects editing; final mixing; and layback, all of which are described in the sections that follow.

Layover

In the layover process, the edited master video, time code and edited sound track are simultaneously transferred to a ¾-inch videocassette recorder and multitrack (16- or 24-track) audio recorder. To control and synchonize the sound post-production process, both the videocassette and multitrack audiotape need to contain matching time code numbers from the edited master tape.

The time code should be recorded on the last track of the multitrack audio tape. On a 24-track tape, for example, the time code should be placed on track 24. Track 23 would be left blank, to provide a buffer that will help reduce the possibility of crosstalk interference between the time code and sound information stored on the other tracks.

The two dialog tracks from the edited master tape should be recorded on two separate tracks, and one track should be reserved for recording the final mix. On a 24-track recorder, this leaves a maximum of 19 tracks for use in the sweetening session. If more tracks are required for the project, some tracks can be preduped, or an additional multitrack recorder can be used.

Projects using the PAP layover process, a process designed for film sound post-production, should keep some important considerations in mind. First, the cut dialog tracks must be transferred to the multitrack recorder at the same time that the film workprint picture is transferred to the ¾-inch videocassette recorder, in order to guarantee *perfect* synchronization between the picture time code stored on the videocassette and the multitrack time code. This is the "double system" transfer method described in Chapter 2.

In the PAP process, the multitrack recorder can also be locked electronically to "rock and roll" dubbing projectors—projectors designed to run the film workprint smoothly in forward or reverse. When run synchronously with sets of 35mm dummies—machines dedicated to playing back 35mm magnetic sound tracks—the special sound effects can be edited on the multitrack audio recorder, while the dialog or music can be cut on separate 35mm magnetic tracks.

During all types of electronic sound editing, I recommend using regenerated time code—time code that has been processed so the digital pulses possess their original crisp, sharp edges. By rerecording with time code that has not been regenerated, you risk losing sync during the recording process.

At the end of the layover process, the ¾-inch videocassette will contain the edited master picture with time code displayed in a "window," the two edited sound tracks and matching time code recorded on the videocassette address track. With the completed cassette in hand, you are ready for the spotting session, the next stage of electronic sound post-production.

Spotting

The purpose of the spotting session is to locate and log the time code numbers for any dialog splits, music scoring, sync effects, background fills, automatic dialog replacements (ADR) or any other special sound work that must be performed during electronic post-production. Usually, the first step is to schedule a screening of the ¾-inch tape created in the layover session. All the production personnel responsible for deciding on and performing this sound work should be present at the screening, and the cassette should be played on a machine that is capable of still-frame and controlled frame-by-frame tape advance.

If scheduling one screening session with all the required personnel is not possible, you can make and distribute copies of the layover cassette. Just make sure that all the "spotters" understand their job: to log accurate time code numbers for the start times and length of all special sound elements. When the spotting process is complete, all of the time code numbers are either hand logged or, in the case of the PAP system, stored on a floppy disk. This list is then "read into" the electronic sound post-production system.

Sound Effects Editing

The sound effects editing or "pre-lay" session is the stage where all synchronous narration, music, background fill, audience reaction and sound effects tracks are added onto separate tracks of the audiotape generated in the layover process. First, crew members review the list of special sound elements generated in the spotting session, and they assign a track to each element on the list.

Before the editing actually begins, the sound mixer should make sure that all of the necessary ¼-inch or multitrack mixdown elements derived during automatic dialog replacement, Foley (synchronized sound effects) and scoring sessions are on hand. The pre-lay process itself is usually performed by one or two crew members located in a room that is separate from the main mixing bay, to avoid tying up extra personnel and the main mixing board. The pre-lay room is normally equipped with a smaller mixing board, a 16- or 24-track audio recorder, several cartridge decks for playing special effects, and ¼-inch and ½-inch audiotape machines. If the PAP process is being used, the pre-lay room will also have several 35mm dummy projectors.

Once the session begins, the crew member in charge watches the videocassette picture in sync with the 24- or 16-track sound recording, previewing and recording each element on the list on its assigned track. If the PAP system is being used, the pre-lay supervisor can also perform computerized crossfades, sync adjustments of as little as 1/10 of a frame, controlled spot erasure "scraping" and controlled dialog track splitting.

Final Mixing

In the final mixing stage, the pre-layed tracks are combined, equalized and mixed into either monaural or stereo sound tracks. The materials used in the final mix include the ¾-inch videocassette copy of the edited master picture and the 16- or 24-track audiotape that contains all of the pre-layed sound. For this session, it's back to the big mixing board, where the talents of the sound mixer will be used to their fullest.

At this stage, there are several mixing options, depending on the delivery requirements for the production. For example, the final mixdown can produce a single monaural track, a pair of stereo tracks, a dialog track that is separated from the music and sound effects tracks (useful for foreign distribution) or music tracks that are separated from the dialog and sound effects tracks (useful when the music must be replaced for home video sales or some other special distribution deal). Although replacing or slipping tracks is still possible at this stage, I recommend performing these functions during the pre-lay stage, to save time and potential trouble during final mixing.

I should sound one note of caution for projects that will be broadcast or played back through a conventional TV set. Although monitoring a final audio mixdown on large concert speakers can make for nice sound in the audio room, it will not give you a very good idea of what the mix will sound like to TV viewers. Since the final product will eventually be heard through the 3- or 4-inch speakers found on most television sets, most experienced producers will listen to the final mix on a similarly small speaker before signing off on the finished mixdown.

Layback

Layback is the final stage in the electronic sound post-production process, the stage in which the finished "mixdown" tracks are synchronized with and recorded onto—or "layed over"—the edited master video recording. Before layback begins, the final mixdown is located on one or more tracks of the same multitrack audiotape that was recorded during the earlier stages of the electronic sound post-production process. As a result, layover is usually the fairly straightforward process of using the SMPTE time code on the multitrack recording to roll and synchronize the audio and videotape recorders, then recording the final mixed sound tracks onto the edited master tape, replacing the old audio tracks.

At the end of the layback process, the edited master recording should always be replayed and monitored, to ensure that the transfer of the sweetened tracks was successful. It is not uncommon for a faulty patchcord or some other seemingly minor malfunction to add distortion to the recording—distortion that could prove very troublesome if it is not discovered before you leave the post-production facility.

CONCLUSION

This review of online conformation and sound sweetening is the final stop on our tour of the electronic post-production process. Looking back over earlier chapters, it is important to remember that electronic post-production is indeed a process—a logical sequence of steps and stages that, like any manufacturing process, produces a final product. With careful preparation and attention to detail, and with a post-production team that knows how to work together, the final product is usually a finished, professional production completed on time and under budget.

Throughout this book, I have concentrated on describing the stages of the electronic post-production process and the various creative, scheduling and budgetary considerations that determine whether "going electronic" is the correct choice for different types of film production projects. I have not, however, delved very deeply into the technical issues surrounding electronic posting. Readers who want or need more technical information should turn to my book *Video Editing and Post-Production: A Professional Guide* (see the Bibliography).

Although electronic post-production may not be the right choice for all film projects, it is becoming an increasingly viable option on a growing variety of productions. Leading filmmakers such as Francis Ford Coppola and George Lucas are becoming more involved in electronic posting, and technical advances have given filmmakers many more options and much more control during electronic post-production.

Given these creative and technological trends, the future for electronic post-production looks very bright. Ultimately, though, it will be the interest and enthusiasm of creative filmmakers and producers that will make the difference, determining whether electronic post-production moves beyond the novelty stage to become a fully viable post-production option.

Appendix A: Reading Edit Decision Lists

To understand electronic post-production, film professionals must have a basic, working knowledge of the edit decision list (EDL). At first glance, a printout of an EDL might seem like a confusing array of numbers and letters (see Figure A.1). However, by breaking an EDL into its basic components, you can see that each set of characters performs a specific function.

The EDL shown in Figure A.1 is written in the American Standard Code for Information Interchange (ASCII) format. Under this format, the EDL contains 10 columns of information, called "fields." The fields include the following information:

1. Field 1 contains the edit number, listed on the EDL in sequential order.
2. Field 2 indicates the source reel for the edit.
3. Field 3 indicates the edit mode (V indicates video only, A indicates audio only, and B indicates both audio and video).
4. Field 4 designates the edit type (C for cut, D for dissolve transition, W for wipe transition, and K for title key transition).
5. Field 5 designates the length, in TV frames, of the transition indicated in field #4. If the transition is a cut, this column would be empty.
6. Field 6 indicates the first frame of the source material, listed in SMPTE time code.
7. Field 7 indicates the last frame of the source material, listed in SMPTE time code.
8. Field 8 designates the location on the edited master tape where the first frame of the source material will begin recording, listed in SMPTE time code.
9. Field 9 indicates the location on the master tape where the last source frame will stop recording. This field can also indicate the duration of the edit, rather than this "out" point.

Figure A.1: An Edit Decision List Showing the 10 ASCII Data Fields

1	2	3	4	5	6	7	8	9
0001	BLK	B	C		00:00:00;00	00:00:00;00	01:00:00;00	01:00:00;00
0001	20	B	D	030	19:37:09;10	19:37:27;13	01:00:00;00	01:00:18;03
0002	11	A	C		17:28:56;22	17:28:57;22	01:00:06;08	01:00:07;08
0003	BLK	V	C		00:00:00;00	00:00:00;00	01:00:00;00	01:00:00;00
0003	22	V	D	030	19:37:09;10	19:37:20;13	01:00:00;00	01:00:11;03
0004	11	B	C		17:29:04;10	17:29:08;14	01:00:14;07	01:00:18;11
0005	14	B	C		17:29:06;21	17:29:12;09	01:00:16;18	01:00:22;06
0006	21	A	C		19:37:27;01	19:37:28;25	01:00:19;24	01:00:21;18
0007	12	V	C		17:29:09;20	17:29:13;12	01:00:19;21	01:00:23;13
0008	10	B	C		17:29:12;04	17:29:26;07	01:00:21;27	01:00:36;00
0009	12	V	C		17:29:15;03	17:29:26;07	01:00:24;26	01:00:36;00
0010	21	A	C		19:37:44;19	19:37:47;07	01:00:33;23	01:00:36;11
0011	BLK	A	C		00:00:00;00	00:00:01;24	01:00:35;03	01:00:36;27
0012	24	B	C		22:42:12;05	22:42:17;27	01:00:34;09	01:00:40;01

10. Field 10 contains the carriage return and line feed information needed for the computer to begin reading the next edit in the list.

Because the carriage return and line feed information is the same for each edit, it is not actually shown on the EDL printout.

As this brief overview indicates, each line on the EDL provides the computerized editing controller with all of the information it needs to perform an edit. Once the controller completes one edit, it simply moves to the next line.

Appendix B: Edit List Management Tools

During the offline editing process, edits are continually changed and recorded over. Consequently, the final offline edit decision list (EDL) will contain a number of edits that have become obsolete. In electronic post-production, this is called a "dirty" list. Previously, it was necessary to spend several hours "cleaning" dirty lists. Overlapping edits had to be trimmed back and undesired edits deleted—a very detailed, time-consuming process.

But that was before. Today, most computerized editing systems feature something called "list management modes," which help editors clean up their dirty lists. There are also several add-on computer programs that help automate the cleaning process. In this appendix, I describe two examples of these list management tools—the 409 program and the trace program.

THE 409 PROGRAM

Using the 409 program, editors can clean edit decision lists in a fraction of the time required with other methods of list management. The program works like this. A dirty EDL, like the one shown in Figure B.1, is entered into the computer's memory from either punch tape or floppy disk—just as long as it was generated on an ASCII-compatible system. In fact, the 409 program can clean several EDLs simultaneously, as long as the lists are entered into the computer in record-time sequential order.

After the dirty list has been loaded into the computer, the editor activates the 409 program by pressing a single key on the editing system's computer keyboard. Pressing the C ("clean") key will eliminate any overlaps in edit-out points. Press-

Figure B.1: A "Dirty" Edit Decision List

```
0001 BLK    B C         00:00:00;00  00:00:00;00  01:00:00;00  01:00:00;00
0001 20     B D    025  19:43:20;19  19:43:26;25  01:00:00;00  01:00:06;06

0002 24     B C         19:50:35;06  19:50:38;06  01:00:05;06  01:00:08;06

0003 22     B C         21:47:17;11  21:47:20;19  01:00:08;00  01:00:11;08

0004 23     B C         19:41:32;12  19:41:42;12  01:00:10;28  01:00:20;28

0005 24     V C         19:50:38;06  19:50:40;06  01:00:10;13  01:00:12;13

0006 21     B C         12:19:55;23  12:20:06;05  01:00:20;13  01:00:30;25

0007 22     A C         22:35:21;10  22:35:24;10  01:00:28;25  01:00:31;25

0008 22     B C         22:35:23;00  22:35:26;00  01:00:30;15  01:00:33;15

0009 21     B C         12:20:21;10  12:20:34;10  01:00:32;25  01:00:45;25

0010 21     B C         12:20:29;10  12:20:32;10  01:00:10;13  01:00:13;13
0010 BLK    B D    025  00:00:00;00  00:00:32;27  01:00:13;13  01:00:46;10
```

ing the J ("join") key will join consecutive audio or video edits that are in sync and that have the same reel numbers. Finally, punching the A ("all") key will automatically activate all phases of the cleaning process. As you might suspect, the A key is used much more often than the other two.

The end product of the 409 process is a clean edit decision list—a list that can be used for online automatic assembly in either the sequential A mode or the checkerboard B mode, both of which were described in Chapter 6. Figure B.2 shows the cleaned version of the EDL from Figure B.1.

The 409 program is both easy to operate and extremely versatile. Over the years, on the basis of extensive practical experience, the program has been preset to accommodate certain fairly common conditions. For instance, the 409 program now assumes the following:

- The EDL is recorded on a floppy disk.
- The floppy list is loaded into disk drive number 1.
- The cleaned list will be returned to a floppy disk.
- "Zero-duration" dissolves should be left in the clean list. (Zero-duration dissolves are sometimes used during offline editing to cut in two adjacent shots in a single edit. During online editing, the zero-duration dissolves are converted to cuts for automatic assembly.)

Of course, the editor can easily change these preset parameters to adapt the 409 program to the conditions present in any given editing situation.

Other features of the 409 program include the ability to take a dual audio channel EDL prepared in a format compatible with one editing system, to clean the list, and to transform the cleaned list into a format compatible with a different editing system. For example, a dual channel audio EDL prepared on a Convergence system could be cleaned and changed into a list compatible with a CMX system. Using the 409 program, an editor can also

- delete all edits other than track-1 audio edits and change the cleaned list into another format
- delete all edits other than track-2 audio edits and change the cleaned list into another format
- delete all audio edits in the clean list
- erase all data from a disk, so the disk can be reused
- combine audio and video edits to form split edits (allowing automatic assembly of those edits in a single event)

The 409 program also features a number of other functions, with still more being added as the program is modified to meet editors' evolving needs.

THE TRACE PROGRAM

The trace program is another effective edit list management tool. Using the trace program, an editor can make several generations of changes to a project, enter the information from the various generations into the computer, and end up with a final EDL that contains time-code references matched to the original source reels.

For example, assume that you're an editor preparing for the second cut of a large project. You certainly aren't going to go back and remake every edit on the list, and you can't splice the videotape to open a scene. Instead, you should use the rough-cut edited master as another source reel, assign it a reel number and use it to edit a second master reel.

In other words, to create a second-cut master, you would edit down a generation. This means that you would be using both the original playback material and the first-cut master to put together the second-cut master. For a third cut, follow the same procedure. As you've probably surmised, the third cut would contain material from the original master, the first-cut master and the second-cut master.

Figure B.2: The Edit Decision List from Figure B.1 After Cleaning with the 409 Program

```
0001 BLK     B C          00:00:00;00  00:00:00;00  01:00:00;00  01:00:00;00
0001 20      B D      025 19:43:20;19  19:43:25;25  01:00:00;00  01:00:05;06

0002 24      B C          19:50:35;06  19:50:38;00  01:00:05;06  01:00:08;00

0Q03 22      B C          21:47:17;11  21:47:19;24  01:00:08;00  01:00:10;13

0004 21      B C          12:20:29;10  12:20:32;10  01:00:10;13  01:00:13;13
0004 BLK     B D      025 00:00:00;00  00:00:32;27  01:00:13;13  01:00:46;10
```

The key to this "editing down" process is to assign each cut master a reel number—a number that was not used on any of the original production reels or any of the other edited reels.

Once the final cut is approved, you will find yourself with an edit decision list that is composed, for the most part, of time code numbers from the recut master(s). Of course, to edit online with the original production tapes, you will need to get back to the time code on those original tapes. In a complicated project with multiple dissolves, audio mixes and other effects, finding those original time code numbers could take days.

That's where the trace program comes in. Using it, an editor can find time-code numbers referenced to the original master tapes almost instantly. The editor begins by loading the EDLs from each cut (generation) into the editing computer. Each list should be assigned the same numbers as the workprint reel to which it corresponds, and, preferably, the lists should be cleaned using the 409 program before they are entered into the computer. Although the computer will ask for the

Figure B.3: The Trace Program

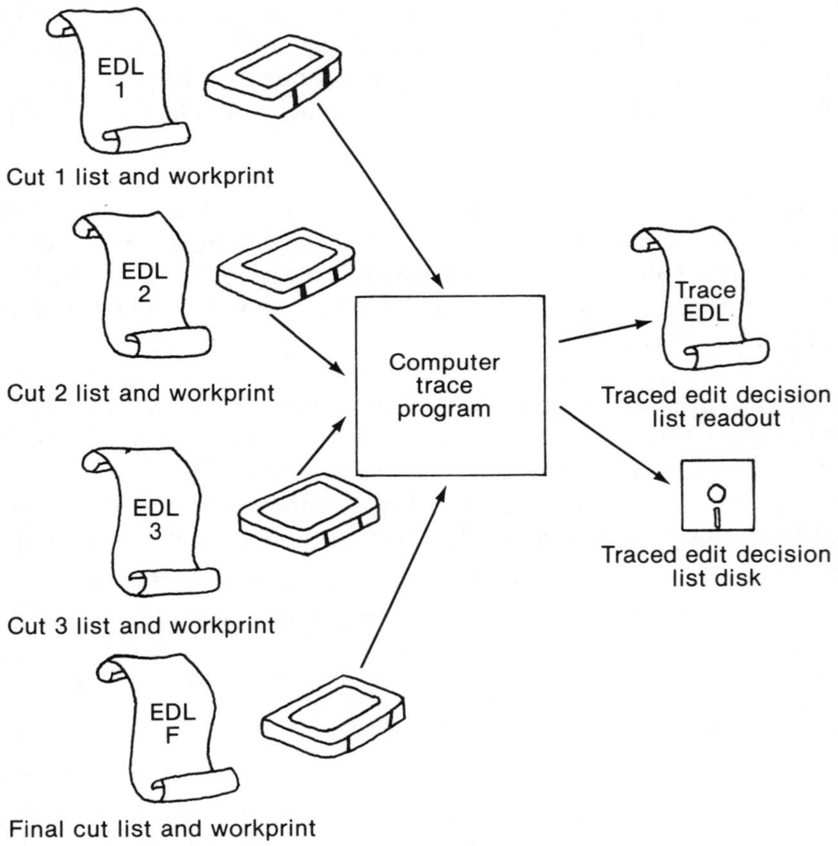

EDL 1

Cut 1 list and workprint

EDL 2

Cut 2 list and workprint

EDL 3

Cut 3 list and workprint

EDL F

Final cut list and workprint

Computer trace program

Trace EDL

Traced edit decision list readout

Traced edit decision list disk

reel number of each list, the lists do not need to be loaded in any particular order or sequence, except that the final version EDL must be loaded as reel "F" (final).

Once the lists have been loaded into the computer, the editor simply presses the "L" (for "list") key on the computer keyboard. In a matter of seconds, the computer will transmit the final, fresh EDL to the line printer. A check of the new final (or "F") list readout will show that all play-in/play-out times refer back to the original master tapes and that all record start times are calculated to correspond to the final version list (see Figure B.3). By pressing another button, the editor can load the new list onto punch tape or floppy disk—whichever is appropriate for the editing situation at hand.

Glossary

Note: Words that appear in italics within definitions are defined elsewhere in the glossary.

A-B roll: Using alternating scenes from two different film or videotape reels to perform dissolves or other image transitions.

ADR (Automatic dialog replacement): The process whereby actors recreate synchronized dialog as they view a playback of their performance.

A mode: Linear method of assembling edited footage. In A mode assembly, the editing system performs edits in a numerical sequence, stopping whenever the *edit decision list* calls for a reel that is not assigned to a VTR. Compare with *B mode*.

answer print: The first color-timed print ordered after the original negative is conformed. See *conformation cutting list*.

ASCII (American Standard Code for Information Interchange): The standard that governs the sequence of binary digits in a computerized *time code* or video editing system.

assemble edit: Electronic edit that replaces all previously recorded material with new audio and video and a new *control track,* starting at the edit-in point. Compare to *insert edit*.

ATR: Audiotape recorder.

audio sweetening: See *sweetening process*.

auto assembly: Process of assembling an edited videotape on a computerized editing system, under the control of an *edit decision list*.

A/V edit: Edit that records new audio and video tracks. Also called *both cut*.

B mode: "Checkerboard" or nonsequential method of *auto assembly*. In B mode assembly, the computerized editing system performs all edits from the reels that are currently assigned to VTRs, leaving gaps that will be filled by material from subsequent reels. Compare with *A mode*.

both cut: See *A/V edit*.

cathode ray tube (CRT): TV picture tube.

character: Single letter, number or symbol used to represent information in a computer or video program.

character generator: Electronic device that generates letters, numbers or symbols for use in video titles. Also a device that converts electronic *time code* into visible numbers displayed on a TV monitor.

checkerboard assembly: See *B mode*.

chroma: Video color.

chroma key: Method of inserting an object from one camera's picture into the scene of another camera's picture by using a solid primary color background behind the object and processing the signals through a special effects generator.

colorimetry: Adjustment of primary and secondary colors without affecting the overall balance of the color signal.

color reversal internegative (CRI): A print of the original negative produced on reversal film, resulting in a negative image.

component video: Video signal in which the luminance and *sync* information are recorded separately from the color information.

composite video: Video signal containing both picture and *sync* information.

conformation cutting list: The *edit decision list* presented to the negative cutter for assembly of the final program from the original camera negative.

conforming: Transferring *edit decision list* information created during *offline editing* to allow the final assembly of the finished program in *online editing*.

control track: The portion of the video recording used to control the longitudinal motion of the videotape during playback or recording.

control track editor: Type of editing system that uses frame pulses on the videotape *control track* for reference.

dailies: The production footage exposed during each day's film shoot.

disk: See *floppy disk*.

dissolve: A video transition in which the existing image is partially or totally replaced by superimposing another image.

doubleperf: Film stock with sprocket perforations along both edges.

dub: To make a copy of a video recording.

dump: To copy stored computer information onto an external medium such as hard copy, paper tape or floppy disk.

dupe: To duplicate a videotape; same as *dub*. Also a duplicate copy of a tape.

EBU: European Broadcast Union.

edit decision list (EDL): List of edits performed during *offline editing*. The EDL is stored in *hard copy, floppy disk* or *punch tape* form and is used to direct the final *online editing* assembly of the video programs.

EDL: See *edit decision list*.

electronic editing: Process of assembling a finished video program in which scenes are joined together without physically splicing the tape. Electronic editing requires at least two VTRs: a playback VTR and a record VTR.

electronic scratch pad: Section of a computer editing program used for making calculations.

encoding: Adding technical data such as *time code,* cues or closed-caption information to a video recording.

event: Number assigned by the electronic editing system to each performed edit.

fade: Usually, a *dissolve* from full video to black video or from full audio to no audio.

field: Half of one television *frame*. In *NTSC* video, 262.5 horizontal lines at 59.94 Hz; in *PAL,* 312.5 horizontal lines at 50 Hz.

film chain: A total system, composed of film projector, slide projector and video camera, used to convert film or slides to video. Also called *telecine*.

film-to-tape transfer: The process of recording film images onto videotape, using a *film chain* and VTR combination.

first trial print: See *answer print*.

floppy disk: Flat, flexible magnetic medium used to store data in computer-readable form. In video editing, floppy disks are used to store *edit decision lists*. Compare with *hard copy, punch tape*.

flying spot scanner: System used for transferring film to videotape in which the electron beam inside a *cathode ray tube* continuously scans the moving film. Flying spot scanners have replaced the mechanical pulldown mechanisms used in early *film-to-tape transfers*.

Foley: The process of recording, in sync, sound effects such as hoof beats, footsteps, etc.

frame: One complete video picture, composed of two *fields*.

frame store device: A digital electronic memory device that scans and stores a complete video frame. Frame stores are used in video special effects systems.

full coat: Film stock with a layer of magnetic oxide covering the entire surface.

general purpose interface (GPI): An electronic device containing several electronic switches that can be activated by a remote data signal. In computerized editing systems, GPIs allow the computer to control various remote components.

generation: Copy of original video program material. The original videotaped material is the first generation. A copy of the original is a second generation tape, and so on. Generally, the edited master tape is a second generation tape.

generic tape: Usually, an edited master tape that does not include situation-specific titles or tags. This specific information is added later, when a copy of the generic master is prepared for broadcast in individual markets.

GPI: See *general purpose interface*.

hard copy: Printed version of computer *output*. In video editing, the printed, "human-readable" version of the *edit decision list*. Compare with *floppy disk, punch tape*.

hardware: Mechanical, electrical or magnetic equipment used in video recording or editing.

hertz (Hz): A unit used to measure frequency. One hertz equals one cycle per second.

hi con: Slang term for a high contrast black-and-white print. Hi con prints are commonly used as a masking source for creating video special effects.

initializing: Process of setting the computer edit program to proper operating conditions at the start of the editing session.

in-point: Starting point of an edit.

input: External information that is entered into a computer.

insert edit: Electronic edit in which the *control track* is not replaced during the editing process. The method of electronic editing in which a new segment is inserted into program material already recorded on the videotape. Compare to *assemble edit*.

interlace scanning: *NTSC* television scanning process in which two *fields* of video are interlaced to create one full *frame* of video.

interpositive: A print of the original negative that creates a positive image.

key: Electronic method of inserting graphics over a scene (luminance key) or of placing one video image into another (*chroma* key).

layback: Transferring the sweetened audio track back to the master videotape. See also *sweetening process.*

layover: Transferriing the edited sound track to a multitrack audio recorder for final sound equalizing and mixing.

list management: On computer editing systems, a feature that allows the editor to change, trim or shift editing decisions stored in the editing computer's memory.

load: To transfer data to or from a storage device. See also *input*.

lookahead: Feature of some computer editing systems that allows the editor to precue one VTR to an upcoming edit while another edit is still in progress.

looping: See *ADR*.

magnetic disk: See *floppy disk.*

mag track: Slang term for magnetic film, particularly 35mm film stock containing one or more oxide strips for recording sound information.

monoperf: Film stock containing sprocket perforations along only one edge. Compare with *doubleperf.*

M/E tracks: Music and sound effects tracks used for foreign-language program distribution.

monochrome: Black-and-white.

multiple record: Process of recording more than one edited master simultaneously.

NAB: National Association of Broadcasters.

nagra track: The ¼-inch sound tape used on a Nagra sound recorder.

NTSC (National Television Standards Committee): The group that established the color TV transmission system used in the United States.

NTSC color video standard: The U.S. standard for color TV transmission, calling for 525 lines of information, scanned at a rate of 30 frames per second.

offline editing: Preliminary post-production session, used to create the edited *workprint* and final *edit decision list* for *online editing.*

online editing: The final editing session, in which the edited master tape is assembled from the original production footage, usually under the direction of an *edit decision list*. See also *auto assembly.*

out-point: End point of an edit.

output: Information extracted from a computer.

PAL (phase alternating line): A color TV standard used in many countries. PAL consists of 625 lines scanned at a rate of 25 *frames* per second. Compare with *NTSC color video standard.*

panscan: The programming of a *telecine system* to scan only selected areas of the image when a wide-screen feature is transferred to video.

pickup cassette: A videocassette copy containing only the retransfer of selected film material rather than the entire original transfer.

preblacked videotape: Videotape used for *insert editing* that contains a recorded control track, a black video signal and time code.

pre-lay: The phase of the audio *sweetening* process during which music and sound effects are added to different audio tracks prior to the final mix.

punch tape: Paper tape punched with holes to represent data in computer-readable form. Used as a method of storing *edit decision lists.* Also called paper punch tape or paper tape. Compare with *floppy disk, hard copy.*

raster: Area of the TV picture tube that is scanned by the electron beam. Also, the visual display present on a TV picture screen.

recall: Retrieving a previously performed edit decision from the computer's memory, allowing for corrections or reedits.

resolution: Amount and degree of detail in the video image, measured along both the horizontal and vertical axes.

safe action area: The area that will safely reproduce on most TV screens; 90% of the screen, measured from the center.

safe title area: The area that will produce legible titles on most TV screens; 80% of the screen, measured from the center.

slide chain: See *film chain.*

slipping: Adjusting portions of the sound track in physical relation to the picture.

SMPTE: Society of Motion Picture and Television Engineers.

SMPTE time code: Binary *time code* denoting hours, minutes, seconds and frames.

software: Computer and video programs and their associated documentation. Compare with *hardware.*

split edit: Edit in which the audio and video signals are given separate in- and/or out-points, so the edit takes place with one signal preceding the other.

split production tracks: An editing technique in which dialog sequences are recorded on alternate tracks to accommodate sound mixing and equalizing.

spotting: The process of screening a final workprint to determine the exact points for adding sound effects and music.

sweetening: The process of mixing sound effects, music and announcer sound tracks with the edited master tape's audio track.

sync: Short for synchronization. Usually refers to the synchronization pulses necessary to coordinate the operation of several interconnected video components. When the components are properly synchronized, they are said to be "in sync."

telecine system: An optical/electronic system used for transferring film to videotape. See *film chain*.

textless: The finished, edited program, minus any titles or text screens.

three-stripe: Magnetic film stock containing three separate stripes of magnetic oxide coating.

time code: Electronic indexing method used for editing and timing video programs. Time code denotes hours, minutes, seconds and frames elapsed on a videotape.

traveling mattes: Animated film sequences used as a masking source for creating visual special effects.

trim in/out: Making minor adjustments to edit points. Incoming and outgoing shots may be extended or shortened by designating either a plus (+) or a minus (−) trim on the editing controller.

user bits: Portions of the *SMPTE time code* left blank for adding user information.

video switcher: Electronic equipment used to switch among various video inputs to a record VTR.

videotape: Oxide-coated, plastic-based magnetic tape used for recording video and audio signals.

waveform monitor: A type of test equipment used to display and analyze video signal information.

window dupe: Copy of an original master recording that features character-generated *time code* numbers inserted in the picture. Window dupes are often used in *offline editing.*

wipe: Special effect transition in which a margin or border moves across the screen, wiping out the image of one scene and replacing it with another.

workprint: Edited master recording created during *offline editing.*

Selected Bibliography

Anderson, Gary H. *Video Editing and Post-Production: A Professional Guide*. White Plains, NY: Knowledge Industry Publications, Inc., 1984.

Annegarn, M.J.J.C. and R.N. Jackson. "Compatible Systems for High Quality Television." *SMPTE Journal* 92:7 (July 1983).

Blair, Ian. "Successful Video Coloring." *On Location* (February 1985).

Bogre, Michelle. "The Art of Film to Tape Transfers." *On Location* (January 1985).

Childs, Ian and J. Richard Sanders. "New Capabilities for a Line-Array CCD Telecine." *SMPTE Journal* (December 1983).

Color Timing. Eastman Kodak publication H-1.

Dager, Nick. "Film Post-Production Enters the Electronic Age." *Millimeter* (July 1985).

Duffy, Robert and Joseph Roizen. "A New Approach to Film Editing." *SMPTE Journal* 91:2 (February 1982).

Film/Telecine Colorimetry. Eastman Kodak publication H-62.

Fremont, Ade M. "Digital Video Processing for Telecine." *Communication and Broadcasting* 8:2 (February 1983).

Lightman, Herb A. "Film and Time Code." *On Location* (April 1985).

Making Television Pictures from Films and Slides. Eastman Kodak publication H-40-12.

Markle, Wilson. "Laboratory Techniques for Television Film." *SMPTE Journal* (February 1983).

North, John and Michael Werner. "Electronic Film Post Developing Fast." *Millimeter* (March 1983).

Patterson, Richard and Stephen Potter. "Post Audio Processing: Electronic Production Techniques." *An American Cinematographer* Reprint (1983).

Poetsch, Dieter. "FDL 60—an Advanced Film Scanning System." *SMPTE Journal* 93:3 (March 1984).

Rank Cintel MK III Telecine Handbook, no. C432.

Rank Cintel MK III C Telecine (Digiscan) Operators Handbook.

Rogers, Pauline B. "A Computerized System to Edit Film to Tape." *EITV* (November 1982).

Rooney, Joseph E. "Film and Videotape Editing: The Process of Conformation." *SMPTE Journal* 93:2 (February 1984).

Strong, Michael J. "SMPTE Time and Edit Code and Its Application to Motion Picture Production." *SMPTE Journal* 93:3 (March 1984).

Technical Experience with Datakode. Eastman Kodak publication V3-517.

Techniques of Telecine Video Operation. Videofilm Notes. Eastman Kodak publication H-40-13.

Television Film Editing and Splicing Techniques. Videofilm Notes. Eastman Kodak publication H-40-8.

Using Films for Television. Videofilm Notes. Eastman Kodak publication H-40-11.

Weston, Daniel S. "Film and Video Options for Production." *SMPTE Journal* 92:11 (November 1983).

Weynand, Diana. "Disk Editing Spins into Serious Contention." *On Location* (June 1985).

Index

ABOUT THE AUTHOR

Gary H. Anderson began his videotape editing career at KCOP-TV, an independent station in Los Angeles, and later worked for Trans-American Video and Vidtronics, Inc. He is currently a videotape editor at Unitel Video, Inc., Hollywood, CA. A member of the Academy of Television Arts and Sciences, he serves on the Academy Blue Ribbon panel for judging videotape editing awards. He is also a member of the Society of Motion Picture and Television Engineers and the International Alliance of Television and Stage Employees. Anderson is the author of *Video Editing and Post-Production: A Professional Guide,* published by Knowledge Industry Publications, Inc.

Anderson's editing credits include network series such as, "Don Kirchner's Rock Concert," "Burt Sugarman's Midnight Special," "That's My Mama," "Soap," "Benson" and "Family Ties," as well as numerous pilots, documentaries, musical productions, commercials and music videos. He has won four national Emmy awards and five nominations for outstanding videotape editing, plus a Monitor award nomination for Best Editor—Broadcast Entertainment.

Related Titles from Knowledge Industry Publications, Inc.

PROFESSIONAL VIDEO PRODUCTION
Ingrid Wiegand

ISBN 0-86729-067-6	hardcover	$39.95
ISBN 0-86729-112-5	student edition, 5-copy minimum	$24.95

VIDEO EDITING AND POST-PRODUCTION
A Professional Guide
Gary H. Anderson

ISBN 0-86729-070-6	hardcover	$34.95
ISBN 0-86729-114-1	student edition, 5-copy minimum	$24.95

PORTABLE VIDEO
ENG and EFP
Norman J. Medoff and Tom Tanquary

ISBN 0-86729-147-8	hardcover	$39.95
ISBN 0-86729-148-6	student edition, 5-copy minimum	$24.95

COMPUTERS IN VIDEO PRODUCTION
Lon McQuillin

ISBN 0-86729-182-6	hardcover	$39.95

PROFESSIONAL VIDEO GRAPHIC DESIGN
The Art and Technology
Ben Blank and Mario R. Garcia

ISBN 0-86729-188-5	hardcover	$29.95

TRAINING WITH VIDEO
Steve R. Cartwright

ISBN 0-86729-132-X	hardcover	$36.95

Available from Knowledge Industry Publications, Inc., 701 Westchester Ave., White Plains, NY 10604.

Will electronic post-production save time and money on a project? What are the benefits and drawbacks of electronic post-production? What is actually involved in "going electronic?"

Electronic Post-Production: The Film-to-Video Guide provides makers of commercials, features, network series, broadcast syndication, music videos, and corporate and educational television with answers to these and other questions surrounding the electronic post-production of film material.

It describes the various technical, budgetary and creative considerations involved in determining whether electronic post-production is appropriate for a particular project. It also compares the available options—e.g., edit on film, transfer to video; edit on video, distribute on video; transfer to video, edit on video, conform negative for film release—and it analyzes actual budgetary, time and other factors.